EPISTEMIC ANALYSIS

SYNTHESE LIBRARY

STUDIES IN EPISTEMOLOGY,
LOGIC, METHODOLOGY, AND PHILOSOPHY OF SCIENCE

VOLUME 173

PAUL ZIFF

The University of North Carolina at Chapel Hill, Department of Philosophy

EPISTEMIC ANALYSIS

A Coherence Theory of Knowledge

D. REIDEL PUBLISHING COMPANY

A MEMBER OF THE KLUWER ACADEMIC PUBLISHERS GROUP

DORDRECHT / BOSTON / LANCASTER

BD
161
.Z54
1984

Library of Congress Cataloging in Publication Data

<u>CIP</u>

Ziff, Paul, 1920–
Epistemic analysis.

(Synthese library ; v. 173)
Bibliography: p.
Includes index.
1. Knowledge, Theory of. I. Title.
BD161.Z54 1984   121   84–6986
ISBN 90–277–1751–6

Published by D. Reidel Publishing Company,
P.O. Box 17, 3300 AA Dordrecht, Holland

Sold and distributed in the U.S.A. and Canada
by Kluwer Academic Publishers,
190 Old Derby Street, Hingham, MA 02043, U.S.A.

In all other countries, sold and distributed
by Kluwer Academic Publishers Group,
P.O. Box 322, 3300 AH Dordrecht, Holland

Printed in The Netherlands

TABLE OF CONTENTS

For Charley, Sam, Ben-Cat, Benglew, Tao, Hansel, and for Georgie, Matt, Kate, Andy and Loredana, and for the Lord Osiris.

PREFACE

THIS ESSAY was begun a long time ago, in 1962, when I spent a year in Rome on a Guggenheim Fellowship. That twenty one years were required to complete it is owing both to the character of the theory presented and to my peculiar habits of mind.

The theory presented is a coherence theory of knowledge: the conception of coherence is here dominant and pervasive. But considerations of coherence dictate an attention to details. The fact of the matter is that I get hung up on details: everything must fit, and if it does not, I do not want to proceed.

A second difficulty was that all the epistemological issues seemed too clear. That may sound weird, but that's the way it is. I write philosophy to make things clear to myself. If, rightly or wrongly, I think I know the answer to a question, I can't bring myself to write it down. What happened, in this case, is that I finally became persuaded, in the course of lecturing on epistemology to undergraduates, that not everything was as clear as it should be, that there were gaps in my presentation that were seriously in need of filling. My gap filling and preoccupation with details may seem to some as unnecessary. I can't help that. The aim of philosophic analysis is an ultimate synthesis. Just as one must slice the sun into many small parts and repack it, if it is to fit into one's pocket (according to the splendid Banach-Tarski theorem), so one must approach the enormous sphere of epistemology: it must be sliced into many small pieces and repacked so that it will fit into one's small head. Every piece has to fit, nothing is to be thrown away.

That is not to say, however, that I am concerned to accredit every loose use of the verb 'know' or every curious claim to knowledge. I am prepared to glance at, and then pass by, such utterances as 'I know that my redeemer liveth', 'I know that I am going to be acquitted' and so forth. The charitable view of the matter is to sup-

ix

pose that when people mouth such utterances, they are not making genuine claims to knowledge, and no doubt some of them know that. Not being in favor of epistemic charity, I think it would be better if people ceased their self indulgent abuse of the word 'know'.

I am concerned to articulate a coherent conception of knowledge. A fine philosopher, Paul Henle, once told me that he was convinced that he believed something absurd, this as an induction from the fact that all philosophers in history have had, what later proved to be, absurd beliefs. I don't doubt that I don't know some of the things I think I know: a dreadful thought, but one that one must do one's best at least to stare out of countenance. There are, nonetheless, some things that I do know: it is the task of epistemic analysis to make clear how and why this can be so.

Although this essay is divided into chapters with titles, neither the chapter divisions nor the titles should be taken too seriously. The thought I endeavor to articulate is a book length thought; the chapter divisions are places to pause, the chapter titles merely suggest focal points. (Chapters I through III are, for me, an *Introduction* and a *Rondo Capriccioso*, IV an *Allegro ma non troppo*, V a *Coda*; VI through X an *Andante tempo rubato*, XI an *Allegro con brio*, XII through XV a prolonged *Allegro jocoso*, leading to a brief *Cavatina*.)

I am indebted to many in the preparation of this manuscript: To the Guggenheim Foundation for a grant that enabled me to begin writing it in 1962; to the University of North Carolina Computation Center for the aid supplied in the creation of appropriately formatted data sets. I enjoyed and profited from discussions of epistemological issues with Frank Sibley in Rome in 1962. But I also am indebted to Robert Brandon, Stanley Munsat, Michael Resnik, Jay Rosenberg and Douglas Stalker, each of whom read an earlier draft of this manuscript and made valuable criticisms and suggestions. I am also indebted to James Coley for the preparation of the index. Finally, I am enormously indebted to Loredana Vanzetto Simpson for constant and invaluable assistance over the past two years.

Chapel Hill   PAUL ZIFF

EPISTEMIC ANALYSIS

"Ouy, ou tout vif aller es cieulx"

CHAPTER I

LINGUISTIC PRELIMINARIES

1. EPISTEMIC ANALYSIS, as I conceive of it, is concerned with the analysis of knowledge. The precincts of my concern have, however, been determined by the vagaries and divagations of the words 'knowledge' and 'know' in English, or, more specifically, in Standard North Eastern American English, or, if you will, in my own idiolect. (There are, of course, aspects of my idiolect that make me somewhat suspect as a claimant to the title of a standard speaker of Standard North Eastern American English. But I am aware of these aberrations, capable of recognizing their peculiar manifestations, and, when need be, of avoiding them.)

It is not the words 'know' and 'knowledge', but the conceptions that these words serve to express that provide the principal loci of epistemological research. But our epistemic conceptions are mirrored in our use of epistemic terms, and, in default of any better glass, an epistemologist must appear to be a verbalist.

2. One remarkable feature of the verb 'know' in Standard North Eastern American English, or SNEAE for short, is that it appears to be a single word.

The verb 'know' occurs in a great many distinct and diverse linguistic environments. The situation is further complicated by the fact that the real diversity is not always explicitly marked.

The sentences 'I know George' and 'I know Dante' are alike in that each has a proper noun as the direct object of the verb. If George's last name is 'Dante', then what is said in uttering the sentences may also be much alike. If, however, 'Dante' is used to refer to Dante Alighieri, then what is said in uttering 'I know Dante' is likely to be that I know Dante's works (though it could simply be that I know who Dante was).

One may speak of knowing George, Dante, Rome, Lycidas, Middle C, the rate of inflation, the answer, the purpose and so forth. One also speaks of knowing who someone is, or who did something; of what something is, or what happened; of whether or not something is the case; of how it happened, or how to do something; and so forth. In contrast with the verb 'know', the noun 'knowledge' occurs in a less confusing set of linguistic environments. One speaks of knowledge of or about something, of the knowledge or a knowledge; of trivial knowledge; and so forth.

3. The first problem one must cope with in attempting an epistemic analysis of 'know' is this: is one, in fact, dealing with only one word? An easy approach is provided by standard dictionary practice: diversity of syntactic categorization is the primary factor. 'Know', in 'He is in the know', is standardly classed a different word than 'know', in 'I know George'. But being in the know is simply a matter of having certain knowledge.

A second factor is radical diversity of meaning. Thus 'mood', as in 'He was in a bad mood', and 'mood', as in 'The mood of the syllogism', are accounted two different words. Although it is difficult to argue the matter at this stage of the analysis, there doesn't seem to be any reason to suppose that such diversity is to be found in connection with 'know'. One doesn't easily infer anything from the mood of a person to the mood of a syllogism, but there are easy inferences between all the various uses of 'know'. Thus knowing George may be a matter of knowing who he is, what he does, that he doesn't lie, how he behaves and so forth. So one may say such things as 'If you knew George, you'd know that he doesn't do that sort of thing'.[1]

4. From a comparative linguistic point of view, SNEAE appears to be oddly at odds with such languages as Italian, French, Spanish, German, Polish, Hindi, Tagalog, to mention only the most prominent.

In each of these languages, one finds two distinct words that correspond loosely to the uses of 'know' in the two sentences 'I know George' and 'I know that George is at home'.

5. That languages differ is not surprising, or news. When we talk about sheep, in SNEAE we make do, by and large, with 'ram', 'lamb', 'ewe' and 'sheep'; Arabians seem to need fifty for our four. Eskimoes have at least twelve words in place of our single 'snow', whereas Aztecs dispense with the niceties of our distinctions between 'snow', 'slush', 'ice' and so forth.

One need not be hard put to explain why languages differ in these respects. The morphology of a language is a function of the demands of an ecosystem. Dwellers along the northeastern seaboard of the United States, where SNEAE is spoken, have few encounters with sheep: mouflon and urials do not stroll the streets of New York. The differences between Arabic and SNEAE in ovine terminology are explicable and predictable. But the differences between, say, Italian and SNEAE in epistemic terminology, though certainly explicable, warrant attention.

6. The same distinctions may be made in different languages in different ways. Gilbert Ryle, in *The Concept of Mind*, discussed the differences between what he called "task" verbs and "achievement" verbs.[2] The verb 'know, in 'I have known George for ten years' is an achievement verb, and that it is is indicated by the fact that 'know' has no present continuous form in standard English English, or in SNEAE for that matter. ('I am knowing George for ten years' does occur in the New York area, but it marks a Yiddish English dialect.) The distinction between task and achievement verbs, though evidenced syntactically, is not explicitly marked in English syntax. Such a distinction is explicitly marked in Slavic languages by so-called "verb aspects". In consequence, Ryle's insightful remarks about the distinction between task and achievement verbs would, in Russian translation, appear somewhat banal. (They would be akin to

pointing out, in English, that if one says 'He was savaged by dogs', one is speaking of the past, whereas if one says 'He will be savaged by dogs', one is speaking of the future.)

In Italian, one translates the English 'I know George' by 'Conosco Giorgio', 'I know how to play tennis' by 'So giocare a tennis', and 'I know that George is at home' by 'So che Giorgio è a casa'. 'Knowledge' is standardly translated as 'conoscenza', though sometimes as 'sapienza', although 'to the best of my knowledge' could be rendered by 'per quanto io sappia', while 'to his knowledge' by 'a sua conoscenza'.

The contrast between 'conoscere' and 'sapere' in Italian (or between 'connaître' and 'savoir' in French, 'kennen' and 'wissen' in German, and so forth) is not explicit in English: that it is implicit can hardly be doubted. Not having the ovine concerns of the Arabs, it is not surprising that speakers of SNEAE have a meagre ovine terminology. But one can hardly suppose that speakers of English lack the epistemological concerns of the rest of the civilized world.

7. The use of 'know' that corresponds to that of 'conoscere' and the use that corresponds to that of 'sapere' are not explicitly marked in English by distinct phonological forms, but they are syntactically indicated. When 'know' takes a direct object, as in 'I know George', the use corresponds fairly well to that of 'conoscere'; whereas when 'know' is followed by a sentence nominal, as in 'I know that George is at home', the use corresponds fairly well to that of 'sapere'.

The striking difference between English and Italian is to be found, not in connection with 'know', but with 'knowledge'. 'Conoscenza' is an evident nominalization of the verb 'conoscere', while 'knowledge' is an evident nominalization of the verb 'know'; but 'knowledge' has not been subject to the same pressures as 'conoscenza': 'una persona con una grande conoscenza di matematica' is a person with a great knowledge of mathematics, but 'essere felice di fare la sua conoscenza' is to be happy to meet him. In English, though perhaps not in SNEAE, one says (or said at one time) 'He had carnal knowledge of the woman'. Even so, one need not conclude that Don Juan and Casanova must have been men of great knowledge; and that the Dean of the College knows the members of the faculty does not warrant the claim that the Dean is a man of some knowledge.

8. If one views English from the standpoint of Italian, it might seem as though the verb 'know' must be polysemous: it must have two distinct meanings, one corresponding to 'conoscere' and one corresponding to 'sapere'. But that would be a mistake.

To say 'I know how to play tennis' in Italian, one says 'So giocare a tennis', but to say 'I know how to play tennis' in German, one says 'Ich kann tennis spielen', thus one uses neither 'kennen' nor 'wissen' but 'können'. Should one conclude that 'know' has, then, not two, but three distinct meanings? That, in turn, would suggest that 'sapere' must then have at least two distinct meanings. But in both English and Italian there is an explicit distinction between 'I know how to play tennis' and 'I can play tennis'; a crippled tennis player can say 'I know how to but I cannot play'. Such a distinction is not explicit in German. Should one conclude that 'können' must have two distinct meanings, one of which corresponds to 'know', and one of which corresponds to 'can', for one can, in German, characterize the plight of the crippled tennis player?

In some languages, instead of drawing a distinction between one or more, as one does in English with the contrast between 'a book' and 'books', a distinction is drawn between one, two or more. If one were to view the plural morpheme '-s' from the standpoint of such a language, it might seem as though '-s' had one meaning in the sentence 'There are books on the table' when two books were in question, and another meaning when more than two were in question. But the obvious fact of the matter is that the plural morpheme '-s' has a single meaning in English, and the difference in meaning between the phrases 'two books' and 'three books' is attributable, not to the morpheme '-s', but to the difference between 'two' and 'three'.

The fact that the uses of 'know' that correspond to those of 'conoscere', and those that correspond to the uses of 'sapere', are syntactically marked suggests that the differences in meaning between 'I know the man' and 'I know that the man is at home' are attributable, not to a difference in the meaning of the word 'know' in each case, but to the differences between the meaning of 'the man' and 'that the man is at home'.

9. Comparative linguistic evidence suggests that the verb 'know' is best conceived of as a single word, with a single fundamental meaning, but a variety of senses.

'Conoscere' and 'sapere' are distinct words of Italian primarily because of their distinct phonological forms. But a phonological distinction does not entail a semantic distinction: 'perhaps' and 'maybe' are as good a pair of synonyms as one can ask for, despite their phonological differences. There is no need to confuse the meaning of a word with any of the various senses in which it may be used. Roughly speaking, the meaning of a word is constituted by a set of conditions associated with the word, whereas a sense of the word is a subset of that set. (More precisely, to use a word in a specific sense is to invoke a subset of a superset of the set of conditions associated with the word.)[3]

(An example may be of some service here. 'Division' in 'The division is marching' and 'division' in 'His arithmetic division is incorrect' exemplify one word with two different meanings. But 'brother' in 'She has been a brother to me' and 'brother' in 'He is not literally my brother' exemplify one word with two different senses, not two different meanings. The difference is that although both words have distinct sets of conditions associated with them, say $\{d1\}$ and $\{d2\}$ with 'division' and $\{b1\}$ and $\{b2\}$ with 'brother', the word 'division', apart from puns and so forth, does not supply us with cases in which the associated set of conditions is equivalent to the union of the sets $\{d1\}$ and $\{d2\}$: in contrast, consider the sense of 'brother' in 'I wish I had a real brother.')

'Sense', in English, is cognate with 'senso' in Italian, and 'sens' in

French. An Italian sign, 'senso unico', indicates a one way street. When, in logic, one speaks of "the sense of a relation", one is concerned with the direction of the relation. One uses a word in a special sense when one invokes certain conditions associated with the word to focus the attention of the hearer in a certain direction.

One can think of the meaning of a word as a peculiar sort of variable polyhedron, where each face corresponds to one of the conditions associated with the word, but where the number of faces is subject to variation, the variation being attributable to the intervention of tropes, discourse operators and so forth. When one uses the word, one turns the polyhedron in one direction or another and thus displays various faces. The faces displayed then constitute the sense of the word.

To say that the verb 'know' is used in various senses in English is to state the obvious: it is precisely what is to be expected of any verb of such evident utility. To thread our way through this maze of senses is the problem; but we can, for we shall find that this maze has a plan.

CHAPTER II

ACTIVES AND PASSIVES

10. THE SYNTACTIC, semantic and pragmatic features of epistemic terms are complex.

If I say 'I believe George', I am likely to be saying that I believe what George says; whereas if I say 'I know George', I am not apt to be saying that I know what George says. If, seeing Bjorn Borg standing on a corner, I were to accost him saying 'I know you', I am likely to be saying that I know who he is; whereas if I turn to a friend with me and say 'Look, there's Bjorn Borg: I know him', I may be saying that I am acquainted with him, not simply that I know who he is, though that possibility has not been excluded. If I know that George is in Yucca Flats, then I may know where George is; but if I know where George is, and George is in Yucca Flats, it does not follow that I know that George is in Yucca Flats. (Thus I might say 'I'll have to take you to him; he's in a small town: I don't know its name, but I can get there and find him'.)

If one attends to the various linguistic environments in which the verb 'know' occurs, one can see that, though on occasion, it occurs in passive constructions, many uses of 'know' seem to resist an agential passive form: I may say 'I know George', but I am not as apt to say 'George is known by me'. (One is much more apt to say 'George is known to me' or 'George is known in the community', but I am not concerned with such so-called "agentless" passives here.) I may say 'I know that it is raining', but I am not apt to say 'That it is raining is known by me'; and though I may know what George did, I definitely would not say 'What George did is known by me'.

I would not suggest that the data here is either blatant or obvious. To hear the discordance of these passive constructions, one must attune one's ear to the language. If one listens, one can hear it.[4] Once one has heard it, the phenomenon demands explanation, and to explain it requires probing matters in detail. As has been said, one can "by indirection find direction out".

11. As almost anyone knows, not every active sentence has a corresponding nondeviant passive; the question is: why?

The relation between actives and passives is generally thought to be of the following form: $NP(1) + V + NP(2) \leftrightarrow NP(2) + be + V + en + by + NP(1)$; (the expression '$NP(1)$' represents the first noun phrase, '$NP(2)$' the second noun phrase, 'V' the verb, '$be + V + en$' the passive form of the verb, 'by' the passive phrase marker, and '$\leftrightarrow$' represents a relation.) If this is a correct representation of the relation between actives and passives then, of course, the sentence 'I am ill' has no corresponding nondeviant passive.

To say that a sentence does not have a corresponding nondeviant passive is not, however, to say that one can not in fact form a passive; of course a sentence such as 'I am ill' poses something of a problem; perhaps 'Ill is been by me' is the corresponding passive; but there are more plausible examples. 'I felt ill' is an obvious case: despite the fact that 'ill' is evidently not a noun phrase, 'Ill was felt by me' is the corresponding passive; but, in my dialect anyway, 'Ill was felt by me' is deviant. So to say that not every active sentence has a corresponding passive, which is what is all too often said, is a deplorably loose way of saying that not every active sentence has a corresponding non-deviant passive. But the question still is: why?

Not to digress, it would be a serious mistake to suppose that the distinction between having a corresponding passive and having a corresponding nondeviant passive is a trivial matter. If one comes across the sentence 'Ill was felt by me' in the context of a poem, it need pose no insuperable problems of understanding; for suppose it were embedded in the discourse 'Grief crawled my spine, Ill was felt by me, Touched with each finger tip': one would then most plausibly suppose 'Ill was felt by me' to be the passive corresponding to 'I felt ill'; were that option not available, the identity of the syntactic structure of the sentence would be in doubt, and, in consequence, what was being said would appear to be incoherent. Consequently, under the impetus of the demands of exegesis, one is likely to construe 'ill' as a noun. (What this amounts to will become clear, or clearer, later.)

12. Consider the sentence 'He realized a profit': such a sentence seems to have no nondeviant passive; 'A profit was realized by him' certainly sounds odd. The same seems to be true of 'He pondered a problem': 'A problem was pondered by him' is the corresponding passive but it doesn't pass, or so it seems. If, however, we begin to tell stories here, to embed these sentences in discourses, then appearances can change, and what once seemed to be the case need no longer seem to be the same. For suppose someone were saying that George and Josef had been speculating, had been trying to make profits on the stock market; and I say 'Well, they're bound to succeed', but you reply 'That's not so clear: a profit was realized by George, but not by Josef'. Is the discourse deviant? The sentence 'A profit was realized by him' sounds remarkably odd to my ear, but the expression 'a profit was realized by George, but not by Josef' sounds ever so much less odd. I will not say that it has the ring of pure silver, but certainly it is less leaden than 'A profit was realized by him' sounded apart from the suggested discourse. And much the same is true of 'A problem was pondered by him'; for consider a discourse in which one speaks of 'a problem pondered by Einstein, Bohr and Fermi, all to no avail'. One is not inclined to object to 'a problem was pondered by him and Einstein, Bohr and Fermi' in the way that one is inclined to object to 'A problem was pondered by him' considered apart from the suggested discourse.

It is, or should be, evident that the phenomenon just noted in connection with 'He realized a profit' and 'He pondered a problem' is not at all peculiar to these particular sentences: to cite a few cases, 'George had a good time', 'I saw a possibility', 'He felt a need', 'We took a walk', 'He was hunting a tiger', 'They needed a rest', 'He slept through the day', 'The man thought about a woman', are all cut from the same cloth: each of the passives corresponding to each of these sentences could easily be embedded in a discourse in which the evident oddity of the passive, considered in isolation from the discourse, would be alleviated or, perhaps, even disappear. For example, 'A good time was had by one and all. What about miserable old Josef? Yes, indeed, a good time was had by Josef and all the rest'.

13. To turn to perhaps more perspicuous cases, consider 'George took a chance', which has the dubious passive 'A chance was taken by George', and contrast this pair with 'George took the chance' and 'The chance was taken by George': evidently the switch from 'a' in the first active to 'the' in the second makes all the difference; for I take it that there is nothing dubious about the second passive, 'The chance was taken by George'. Furthermore, even though 'The man needed a woman' has a dubious passive, 'A woman was needed by the man', no such problem occurs in connection with the active sentence 'A man needed a woman', for the corresponding passive is then 'A woman was needed by a man', which is altogether un-problematic. Furthermore, even though 'George passed the time' has the dubious passive 'The time was passed by George', the passive of 'George passed the time in meditation', which is 'The time was passed by George in meditation', is not at all dubious.

All this, of course, wants explaining: why does the switch from 'a' to 'the' in a predicate make such a difference? Why, if 'a' is in the predicate, does it then help to switch from 'the' to 'a' in the subject? Why does the use of a prepositional phrase in the predicate seem to smooth the path of the passive? The answers to these questions are, in fact, relatively simple. These matters seem complicated only if one insists, perversely, on looking at them in the wrong way.

The relation between actives and passives is generally characterized as a transformation of the form: $NP(1) + V + NP(2) \leftrightarrow NP(2) + be + V + en + by + NP(1)$; if one mistakenly thinks of a transformation as some sort of rule, either for mapping or for generating sentences, one is, of course, at once confronted with the insuperable problem of specifying the domain of its operation; otherwise either one will fail to map or generate various nondeviant sentences, or one will map or generate deviant sentences. Fortunately, there is no good reason to think of a linguistic transformation as a rule, despite current dogmas to the contrary; for, on the contrary, transformations are best thought of as devices, indeed, even instruments, for modulating discourse.

14. Let's look away from language for a moment and consider things and relations: if something, a, stands in a relation, R, to something, b, then it is a truth of logic that b stands in the converse of that relation to a. But if Josef is ill and one reports this fact by saying that Josef is ill, one is not reporting that something, a, namely Josef, stands in a relation, R, namely that of being, to something, b, namely ill; on the contrary, one is attributing to a the characteristic, or attribute or property, C, of being ill; thus one is reporting something perspicuously represented as of the form $[C]\,(a)$. Whereas if one reports that George struck Josef, one is reporting something perspicuously represented as of the form $[R]\,(a, b)$. It is a truth of logic, however, that polyadic predicates can be construed as monadic predicates; one can, accordingly, construe something of the form $[R]\,(a, b)$ as of the form $[R(b)]\,(a)$. In consequence, when one says 'George is hunting tigers', what is being said may be represented either as of the form $[R]\,(a, b)$ if one construes the situation as one in which there are specific tigers being hunted by George, or as of the form $[R(b)]\,(a)$, if one construes the situation as one in which George is tiger hunting, regardless of whether specific tigers are being hunted. The expression 'b', in $[R(b)]\,(a)$, then serves to represent an adverbial object.
And now conjure up two states of affairs: in one, we are inclined to say that George is hunting a unicorn; in the other, we are inclined to say that Josef is stroking a tennis ball. Since we (or, more precisely, many of us, but I shall use 'we' to refer to those of us who do) know

that there are no unicorns, we are not likely to think of George's situation as one that involves a relation between George and a unicorn: there are no unicorns to be in relation to. On the other hand, there are tennis balls to be found in this world and not many of us often have occasion to say that someone is stroking a tennis ball unless there is some tennis ball that he is in fact stroking. But what is essential to realize here is that it is not a fact of grammar that there happen to be no unicorns in this world, even though many of us know that there are no unicorns, and it is not a fact of grammar that when someone is stroking a tennis ball, there is some ball that he is stroking. Indeed, this is not invariably the case: someone watching a tennis player practicing in front of a mirror might be told that the player was practicing his topspin backhand drive. The player might comment: 'I'm stroking the ball with the racquet head kept vertical to the ground throughout the stroke, imparting topspin to the ball simply by hitting from low to high'. To object: 'But there is no ball' would merely be, at best, naive. The fact of the matter is that one can be stroking a tennis ball without there being any tennis ball about to be stroked, just as one can be hunting a unicorn without there being any unicorn to be hunted.

15. If someone says 'George is hunting a unicorn', is he, in so saying, saying that something, *a*, namely George, stands in a relation, *R*, namely hunting, to something, *b*, namely a unicorn? And if someone says 'Josef is stroking a tennis ball', is he, in so saying, saying that something, *a*, namely Josef, stands in a relation, *R*, namely stroking, to something, *b*, namely a tennis ball? The answer to the latter question is: perhaps no; to the former: perhaps yes. I have seen shops in South Philadelphia where a substance labelled "Powdered Unicorn Horn" was on sale. Suppose in the vicinity of such a shop I notice a commotion and ask 'What is going on?' The reply is 'A unicorn is being hunted'. 'By whom?' I ask; 'Old Josef' is the reply. So I think, and say echoically, 'A unicorn is being hunted by old Josef that's absurd!' Yes, that is absurd, but the absurd-ity is not a linguistic absurdity.

In my example, I first introduced the sentence 'A unicorn is being hunted' and then, in the discourse, added 'by old Josef', because, of course, the diachronic development of the discourse more strongly suggested that the situation was to be construed as relational. That various cues, linguistic and nonlinguistic, can guide one in one's construal of a situation is not to be denied; but the cues can be subtle and, that they are cues, may go unnoticed. To say 'George ate a fish' at once indicates that a relation is in question; but to say 'George had a fish' does not: 'had' does not, but 'ate' does, indicate a relation. That this is so is, indeed, evidenced by the fact that the passive of the former, 'A fish was eaten by George', is not deviant, whereas the passive of the latter, 'A fish was had by George', is deviant (or certainly seems so to me). (It should not be surprising, however, that 'had' does not indicate a relation, given that 'had' is a tensed form of the verb 'have', and given the role of 'have' in English. Perhaps it should be stressed that, in saying 'had' does not indicate a relation, I am not denying that if George had a fish then there was a relation between George and a fish. But that there was a relation is indicated, when one says 'George had a fish', not by the word 'had', but by the syntactic construction 'had a fish': no relation is indicated, for example, when one says 'George had a nap', or 'had a good time' and so forth.)

Linguistic cues can guide one in one's construal of a situation; but such cues are, at best, merely *prima facie* indications, and various other factors may intervene and prevail. So, rather than speaking of "cues", it would be more precise here to speak of "vectors".[5] Thus in the context of such a discourse as 'Grief crawled my spine, Ill was felt by me, Touched with each finger tip', despite the fact that 'ill' is not a noun, one construes it as a noun, and so hypostatizes illness, in the interest of coherence.[6] The device of hypostasis is readily available to speakers of a language; but, of course, it is easier to employ such a device when a noun is clearly in evidence: it requires greater art to invoke a passive of 'I felt ill' than to invoke a passive of 'I felt pain', and still less to invoke a passive of 'I felt a pain'. And even though 'had' does not indicate a relation in 'George had a fish', one can force the verb to comply by embedding it in an appropriate discourse: 'Did George have a fish? Yes indeed, a fish was had by George, a gator by Josef and a cooter by Sidney', which, again, may not have the pure ring of silver, but it surpasses the dull chink of lead.

16. Whether an active sentence has a corresponding nondeviant passive depends on whether the most coherent interpretation of the sentence, taking into account all relevant vectors, both linguistic and nonlinguistic, requires one to construe what is in question as relational, rather than nonrelational. And what this means is that whether an active sentence has a corresponding nondeviant passive is not simply a syntactic question: one must also take into account semantic, pragmatic and logical factors.

Consider the sentence 'George thought about a trip to Europe': does what is in question involve a relation between George and a trip to Europe? Given simply the linguistic vectors supplied by the sentence, one is, or anyway, I am, inclined to say that it does not. I am inclined to construe the noun phrase, 'a trip to Europe', as an adverbial object, just as one construes 'tigers' as an adverbial object in 'George is hunting tigers', in which case, of course, one could less ambiguously, and hence more clearly, express what is in question by saying 'George is tiger hunting'. (But this option is not available with 'George thought about a trip to Europe': both 'George about a trip to Europe thought' and 'George a trip to Europe thought about' are deviant (at least in my dialect of American English, in SNEAE.) There appear to be not only syntactic, semantic and pragmatic, but cultural restrictions as well on the placement of an adverbial object before a verb. Thus even though 'George is tiger hunting' is nondeviant, 'George is tiger seeking' is deviant, although 'George is seeking tigers' is not: the difference is that hunting is a recognized social practice; seeking is not.)

Given that one is not inclined to construe what is in question as involving a relation, the passive, 'A trip to Europe was thought about by George', seems deviant. But suppose, however, that George was thinking, not just about a trip to Europe, but about a very specific trip to Europe: a trip first to Rome, then to Orvieto, then to Siena and so on. The passive then seems to pose less of a problem. That it still poses something of a problem is, I think, directly attributable to the fact that 'thought', as well as the verb 'think', does not indicate a relation. For notice that the sentence 'George was considering a trip to Europe' readily accepts the passive 'A trip to Europe was being considered by George'. Furthermore, if one emphasizes the importance of what George is supposed to be considering, as is done in, 'George was seriously considering a trip to Europe', the path of the passive appears to be considerably smoother: 'A trip to Europe was seriously being considered by George'. Still more, if one supplies indications that what is in question is sufficiently specific to be referred to specifically, the passive becomes even more plausible.[7] Thus switching from 'consider' to 'plan', and inserting the adverb 'carefully', there is no problem at all with 'A trip to Europe was being carefully planned by George'.

17. It is, or should be, obvious now why switching from 'a' to 'the' in a predicate renders the passive more plausible, and why the same is true of switching from 'the' to 'a' in the subject, if 'a' is in the predicate. If one is to construe what is in question as relational, then one will have to fix on terms of the relation. Characteristically, in English subject-predicate sentences, subjects refer and predicates characterize. Thus, for example, though one might say 'George is a fellow who lives in town', the sentence 'A fellow who lives in town is George' seems odd in isolation, apart from some discourse; the oddity disappears, however, if one switches from 'a' to 'the' in the subject, as in 'The fellow who lives in town is George'; switching from 'a' to 'the' serves to make it clear that a term is available: 'the' is a stronger vector than 'a' in establishing reference. On the other hand, if 'a' is in the predicate, as in 'The man needed a woman', switching

from 'the' to 'a' in the subject will increase the plausibility of construing the predicative expression 'a woman' as a referring expression. Thus 'A man needed a woman' has the nondeviant passive 'A woman was needed by a man', for it is clear in the active sentence that the initial 'a' serves to mark a referrring expression: in consequence, the plausibility of construing its second occurrence as also marking a referring expression is increased. Reference is again the reason why the path of the passive is eased when one adds the prepositional phrase 'in meditation' to 'George passed the time'. For if George passed the time in meditation, it is then indicated that the time in question has a specific character, and hence, if referred to, may be characterized as that time passed in meditation; so: 'The time was passed by George in meditation'.

18. If, on the most coherent construal of an active sentence, the main verb of that sentence indicates that a relation is in question, then the passive of that verb serves to indicate that the converse of that relation is in question. The most obvious reason for invoking the passive transformation is to shift the focus of attention from the first to the second term of the relation. Thus, in the diachronic development of the discourse, 'George ate a fish. The fish eaten by George was a bass', 'a' shifts to 'the' to identify the reference of the subject of the passive construction as that which was referred to in the initial active sentence.

But the primary reason for employing the passive construction is to establish that a relation is in question. Consider the sentence 'George took a wife': does it have a corresponding nondeviant passive? Yes and no must be the answer: yes, if the wife George took was Josef's wife; no, if George merely married. 'A wife was taken by George' is thus unambiguous, even though 'George took a wife' is not. Again, 'George is hunting a tiger' is ambiguous in a way that 'A tiger is being hunted by George' is not. And still again, 'A woman was needed by a man' is unambiguous in a way that 'A man needed a woman' is not; for the latter could simply be a way of saying that the man had a sexual drive, but that would be a deviant construal of the former: at least an object of the drive is then indicated. The expression 'a woman' in 'A man needed a woman' may be either a direct or an adverbial object, but that option is precluded in the passive construction.[8]

That is the *raison d'être* of the passive construction in English.

19. (It must be emphasised that the preceding account of agential passive sentences in English is just that, an account of such sentences in English: I am not concerned with similar forms in other languages. In discussing passive patterns in Kimbundu, a member of the Bantu dialect group, Givón translates the sentence 'Nzua amumono kwa meme' as 'John was seen by me'.[9] Transliterated, the sentence reads 'John they-him-saw by me', for the prefix 'a' is the plural third-person subject pronoun and the invariant marker of the passive in Kimbundu. But although I would not, indeed could not, question Givón's transliteration, the translation is certainly open to question. The sentence 'John was seen by me' is, unless embedded within an appropriate discourse, or propped up by appropriate contextual factors, certainly deviant in English.

There are various possibilities here. Perhaps the correct translation of some sentences in Kimbundu into English requires one to shift from a passive to an active form. Thus, in this case, the best translation might be 'I saw John'. Or possibly speakers of Kimbundu tend to view situations in relational terms in a way that speakers of English do not. In English, or at any rate, in SNEAE, the following discourse is altogether implausible: 'Hello George! I must tell you this: John was seen by me going into a massage parlor'. One would say either 'I saw John going into a massage parlor', or if it was Josef who witnessed the event, 'Josef saw John going into a massage parlor', or if it is not known who witnessed the event, 'John was seen going into a massage parlor', thus one would employ a modified passive form, a so-called "agentless" passive, which effectively precludes the relational implications of the full passive construction. Of course, appropriate background information could change matters; perhaps one could set the scene in such a way that the passive would seem plausible. But possibly such a discourse would be plausible in Kimbundu without such supplementary background information.

Our concern here is with English (or at least with SNEAE). Although many linguists are prone to deny it, the fact of the matter is that there are all sorts of things that can be said in one language that can't be said in another. If there is a single bone on a table, in English one could speak of "a bone" on the table; whereas if there is more than one bone, one could speak of "bones" on the table. But a speaker of Nootka, an Eskimo language, could, in either case, speak of "hamot" on the table without focusing the hearer's attention on the number of bones in question. The English translation of 'hamot' would have to be 'a bone or bones'. Although such a translation would capture the appropriate conditions, it would serve to structure the attention of the hearer in a radically different way than the source expression in Nootka. The English translation focuses the attention of the hearer on an alternative, but no such focus was associated with the source expression.)[10]

CHAPTER III

REFERENCE

20. THE SENTENCE 'I know George' does not, in my idiolect, have a nondeviant passive. But there are discourses in which the sentence can be embedded and in which the passive is then unobjectionable. If one considers a situation in which a group is in question, the shift to the passive becomes plausible. If one is concerned to have someone identify George, and if one member of the group knows George, one might say 'George is known by at least one member of this group, namely by me'. (It should be noted, however, that, even in the suggested context, the simple sentence 'George is known by me' would not ring true.)

There is, as far as I know, no traditional name for the logical stance adopted when one chooses to construe something in relational terms and thus render the passive construction available. So I shall speak simply of "relational construal".

If one wishes to indicate a relational construal of a situation, when what is in question is one person's knowing another, the simplest way to do it is to invoke the passive. If one says 'I know George', it is clear that two terms are available, the referent of 'I', namely the speaker, and the referent of the proper name 'George', to stand in the relation. That one would not ordinarily reply to the query 'Do you know George?' by saying 'Yes, George is known by me', is simply owing to the fact that we do not ordinarily adopt a relational construal of such a situation.

If one wishes to adopt a relational construal of a situation, when what is in question is someone's knowing that something is the case, it can, of course, be done, but matters become much more complex.

21. Speakers of SNEAE, and of any natural language, readily avail themselves of various syntactic and rhetorical devices.

If someone is looking for George, and I know that George is at home, if asked, I might well say 'I know that George is at home'. I would not, in such a case, say 'That George is at home is known by me'. But, again, suppose the situation were quite different: say a group is being interrogated, and it is thought that at least one member of the group knows that George was at home. Someone might plausibly say 'That George was at home is known by at least one member of this group, namely by me'. (Again, the simpler 'That George was at home is known by me' does not ring true, and for the same reasons.) That the passive construction is available in such a discourse is explicable in terms of the syntactic device of nominalization and the rhetorical device of hypostasis.

But since Metaphysics is the awesome offspring of the lawful wedlock of Nominalization and Hypostatization, we shall have to proceed warily.

22. Nominalization is a process, a linguistic process, by which one creates nouns, or noun phrases, out of nonnouns. It takes many forms.

Consider the sentence 'George is obese'. 'Obese' in 'George is obese' is a predicative adjective. A nominalization of 'obese' yields 'obesity'. A nominalization of 'is obese' yields 'being obese'. A nominalization of 'George is obese' yields 'that George is obese'. Still another nominalization of 'George is obese' yields 'George's being obese'. And another is 'George to be obese', while still another is 'for George to be obese'.

Nominals, the products of the linguistic process of nominalization, function as nouns in discourse, and, hence, occur in the positions that nouns occur in in sentences. So one says such things as 'Obesity is deplorable', 'That George is obese is deplorable', 'George's being obese is deplorable', 'For George to be obese is deplorable', 'I don't want George to be obese', 'I know that George is obese' and so forth. Syntactically speaking, nominalization is a fairly straightforward, though complex, process. But, from a semantic point of view, it is a metaphysical dream.

23. One can, by making the devious moves that philosophers are frequently fond of making, make it seem plausible to suppose that to say 'I know that George is at home' is to indicate a peculiar sort of relation.

To suppose that a relation is in question, one must suppose that certain entities stand in the relation: what are they? The first is easy, for 'I' in 'I know that George is at home' indicates the speaker; the second is more of a problem. The word 'that' in 'I know that George is at home' is simply a syntactic marker: it serves to mark a nominalization of the sentence 'George is at home'. In SNEAE, instead of saying 'I know that George is at home', one could just as well say 'I know George is at home': since the nominalization is evident, one can readily dispense with its syntactic marker. The standard metaphysical move is then to suppose that the nominalized expression "refers" to something. The next move is to suppose that one can't refer to something unless that which is referred to "exists".

24. Can one refer to that which does not and never did exist? Authorities say no. (I leave it to you, and spare me, the task of finding out whom I might be referring to.) That is the received view. It is nonetheless untenable: it constitutes a gratuitous misrepresentation of sensible linguistic practice.

'Despite his manners, Nasty is, when he puts his mind to it, a magnificent player'. Whom am I referring to? Ilie Nastase, the Rumanian tennis player. I accomplish this feat of referral by, among other things, relying on a use of the expression, 'Nasty', embedded as subject of the complex sentence cited. It would be excessively naive to suppose that the matter is managed simply by a use of the nickname, 'Nasty'. That phonological form might accomplish little by itself: witness the (spoken) remark, 'Don't be nasty!', which need not be an injunction not to be Ilie. (I also, and of course, relied on various capacities and so forth of my hearers (or readers), but these are not matters that we need go into for the moment.[11]

Did I also rely on the existence of Nastase? He does exist (or did the last I heard). Here we seem to have a crux of the matter: was my act of referring dependent on the present existence of Nastase? More precisely: was his present existence a *sine qua non* condition for my performance of the act of referring to him as I did? And was it equally a *sine qua non* condition for the expression, 'Nasty', in the context of the cited sentence, to have referred to him as it did? The answer to these questions is clearly yes, but this clear yes can invite a hasty erroneous generalization. For one can readily refer to that which does not presently exist. In support of this easy view, it should suffice to invoke the name of Dante.

25. Consider, not Nastase, but Dante, who, unlike Nastase, doesn't exist: he did, but now doesn't. In some timeless sense he does you say? What's that but a fancy way of saying that he doesn't but did exist? Someone talking (timelessly, he says) says 'Dante exists', and I say he's playing word games.

'The author of *The Divine Comedy* was a fine poet'. ('Is a fine poet' one might, and perhaps would ordinarily, say, but that is a trope, here distracting and out of place: notice that one would not now ordinarily say 'Bill Tilden is a great player'.) 'The author of *The Divine Comedy* was a great poet'. To whom am I referring? Dante Alighieri. But he doesn't exist, even though he once did. That's a fact that one has to face up to in formulating an adequate theory of reference. On almost anyone's account then, to perform the act of referring, it is not necessary that that which one refers to exist at the moment of reference. Just a moment ago I referred to Dante Alighieri, and when I did, he didn't exist: though out of existence, he's not out of reach of reference.

26. Nastase does, and though Dante doesn't, at least he did exist. But there are still more rarified forms of reference. 'Despite his inability

to act, the Prince of Denmark is an heroic figure'. Is anyone being referred to? Hamlet, of course, Prince of Denmark, he for whom the native hue of resolution was sicklied o'er with the pale cast of thought. However, unlike Dante, Hamlet presumably never did exist.

What may bother some, even though it is not, in truth, bothersome, is that they may think there can be no truths about Hamlet, given that he does not and never did really exist. Why they think that I don't know, for there are such truths: truths of a sort, if you like, but truths nonetheless. Isn't it true that Hamlet berated his mother? As we all know, or should know by now, the answer to this is at least 'Yes, but': being a fictional character, he didn't have a real mother to berate, and, of course, he did berate his mother, Gertrude, for he accused her of posting with dexterity to incestuous sheets.

So maybe what some think is this: any truth about Hamlet is not simply a truth about Hamlet, the Prince of Denmark, but ultimately a truth about a play: thus some might maintain that any true statement about the Prince must be paraphrasable in terms of a true statement or statements about the play, or about the play and its author, or his time and so forth. Whether all this is so, I don't know, but, anyway, it's beside the point: proper paraphrase can hardly convert a truth into a nontruth. What the possibility of paraphrase might show is that one might be able to say anything one wants to say about the play, *Hamlet*, without ever using the name, 'Hamlet', or the phrase, 'the Prince of Denmark', to refer to a fictional personage. But that does not in the least show that, when one does use such expressions to refer to a fictional personage, one is not, in fact, doing so. There are truths about the author of *The Divine Comedy*: he was an historical personage. And there are truths about the Prince of Denmark: he is a fictional personage. The striking difference between 'The author of *The Divine Comedy* was a fine poet' and 'The Prince of Denmark is an heroic figure' is that what is referred to in the one case did, whereas what is referred to in the other never did, exist. There is (apart from the use of 'was' and 'is') no other relevant difference.

27. Some might say: Hamlet never did exist, yes, but he does exist
in fiction. So, in some sense, he does exist and, so, is available for
reference. It is true that Hamlet is a figure of fiction, a fictional
character who has, so to speak, been around a long time. So he exists
in fiction. I don't think that that's important, but it is true, and some
insist on it.

Thus John Searle, who urges an "axiom of existence": "Whatever is
referred to must exist" (where 'exist' is to be construed, he says,
"tenselessly") hopes to cope with Santa Claus (*Exorceo te impura
creatura* ...), and Sherlock Holmes and others of their ilk by invoking
their fictional existence.[12] "one can", he says, "refer to them as
fictional characters precisely because they do *exist in fiction*".[13] This,
in italics, which later give way to hyphens, as in "exist-in- fiction".[14]
He adds, "To make this clear we need to distinguish normal real
world talk from parasitic forms of discourse such as fiction, play
acting, etc".[15] He further and later adds: "The axiom of existence
holds across the board: in real world talk one can refer only to what
exists; in fictional talk one can refer to what exists in fiction (plus such
real world things and events as the fictional story incorporates)".[16]
Searle does have an engaging way of making a pass at, but just
missing, truth. For this is charming fol-de-rol and la-de-da: nonsense,
in short. Staring at a woman staring in a mirror, a misogynist mutters,
'Hamlet was right: Vanity, thy name is woman'. Has someone
suddenly slipped into a fictional, play acting, let's pretend mode of
discourse? Our misogynist has managed to refer to Hamlet. If she
was in some "fictional ... mode of discourse",[17] it was only because
she used the name 'Hamlet' and misquoted him: it is not that being
in some special mode of discourse somehow enabled her to refer to
Hamlet.

28. If reference requires existence, then how can merely fictional
existence suffice? And if merely fictional existence counts as existence,
then, indeed, more things exist than have been dreamt of in my
philosophy.

For George, who is a nervous type, has, he professes, finally found a friend, Goredl by name, an owl, albeit imaginary. 'Goredl sits on my shoulder, converses with me' he says. 'I don't see him'. 'Of course not', he replies. 'Goredl is an imaginary owl'. 'Goredl is a fine fellow' I say in soothing tones, dialing softly for George's psychiatrist. Does Goredl exist? 'In George's imagination' some might say.

Nastase exists. Dante doesn't, but did. Hamlet didn't, but does in fiction. Goredl does only in George's imagination. Certainly one can refer to Nastase. To Dante. To Hamlet. And to Goredl too. And one can even, and often does, definitely and cheerfully, refer to that which neither does nor ever did exist in any sense whatever. For the battle of Waterloo was an event, an occurrence, a happening of some interest. And one can, and does, refer to it even though such an act of reference fails to satisfy the Searley "axiom of existence". Battles occur, take place, happen, but don't, nonetheless, exist. That a nuclear holocaust will never exist is not a source of comfort: it may occur. And then, of course, to stir old ashes, one can and does refer to the prime number between 19 and 29, to the number 23: can one also, *ipso facto*, infer 'The prime number 23 exists'? And, finally, what is one to say of 'the spin of the wheel', 'the rate of inflation', 'the cost of living' and 'the myriad matters that one refers to'? Does this myriad exist?

29. "To be", we have been told, "is to be the value of a variable". 'Chairs are' one says, and he is not supposed to have run out of breath, but to be breathing the rarer air of ontology. Contrast: 'There is a man at the front door' with 'There exists a man at the front door'. Why is the latter so odd? 'There is a prime number between 15 and 18' – 'There exists a prime number between 15 and 18': here the oddity vanishes.

Roughly speaking, in English the word 'exists' is used in speaking of an element of a domain.

In the domain of prime numbers, is there an element meeting the condition of occurring between 15 and 18? Do prime numbers exist?

Yes, they are elements in the domain of positive integers. 'Then numbers must exist!' But how does that follow? What domain is in question? Say 'the domain of mathematical objects'. Then numbers exist. 'Then mathematical objects exist!' But, again, how does that follow? What is the relevant domain? Say 'Things that exist' – and here one has come a full circle. One might as well say 'Everything that exists exists'.

'Do myths exist?' Yes, they are important elements in the culture of the tribe of logicians.

30. The ubiquity of reference has served to conceal the real character of the act. Reference is easy, readily accomplished in a trice. So, not reference, but the lack of it is apt to be revealing here.

'Supervisor Josef decided your case. The matter is closed. I can do nothing': so I was told by a government clerk wishing to be rid of me. I later discovered that there was no Supervisor Josef. Did the clerk refer to a Supervisor Josef? No: he pretended to: he created the impression of doing so. But he did not in fact refer to anyone. Neither did the expression 'Supervisor Josef' in fact refer to anyone (at least not the expression token uttered by the clerk). Why not?

Not simply because no such person exists. That is part of the story, but it is not the whole story, which is not to deny that, nonetheless, it certainly is part of it. Consider what I was supposed to believe about the supposed referent of the descriptive name 'Supervisor Josef'. I was to believe that a person who is a supervisor, who is called 'Josef', decided a case: I was to believe he existed, for I was concerned with a real case. I then learned that no such person existed. That means I then had no coherent conception of a referent of the expression 'Supervisor Josef': he doesn't exist and he settled my case? That's absurd.

31. Is all this clear? No, not at all. For there are considerable complications here owing to the evolutionary diachronic character of discourse and correlative conceptual schemes.

Do I have a coherent conception of the referent of the expression 'Supervisor Josef'? At first, yes, and then, no, and then, yes, but when finally yes, one radically different from that first formed on hearing the clerk's words. Then I did not, but now I do, realize that Supervisor Josef is a fiction created and invoked by the clerk to deceive me. Now you and I can talk about Supervisor Josef, and in our discourse the descriptive name could well serve to refer to a fictitious personage. But when I first discovered the clerk's duplicity, there was no genuine reference, for no sufficiently coherent set of conditions was then in effect. (A speaker, in uttering an utterance, may invoke conditions which serve to structure the attention of the hearer; but for this to occur, the invoked set of conditions must, among other things, be reasonably coherent.)

32. Consider George and Goredl: 'Goredl is sitting on my shoulder' George says. 'I can't see him' I say. 'Certainly not' George replies, 'Goredl is invisible'. Does the name 'Goredl' refer to anything? Yes. But Goredl doesn't exist. Of course not: he's an imaginary owl. Indeed, if he did exist, I might not be able to form a coherent conception of him. The only way I know of for the referent of 'Goredl' to be an owl sitting on George's shoulder and also to be invisible is to be imaginary or fictional or the like.
Are there truths about Goredl? Yes. Why not? If there were not, there could be no question of coherence. But there is and there are: that Goredl doesn't exist surely follows from the truth that Goredl is merely an imaginary owl and not a real one. What kind of owl is Goredl? Is he a barn owl? I don't know. One would have to ask George: after all, Goredl is a creature of George's imagination, not mine. It is then a truth about Goredl that he's invisible and that he sits on George's shoulder. There's no problem in forming a reasonably coherent conception of Goredl. (One could intone softly to oneself 'Goredl does exist, at least in the imagination': one could, but there's no need to rely on such a mantra.)

33. That coherence is the critical factor in reference is further evidenced by the fact that it serves to resolve ambiguity of reference. George says 'Julius Caesar refused to cross the Rubicon: he just stood at the water's edge barking his head off'. Possibly George is referring to the ancient Roman and is woefully confused. More plausibly, he's referring to his dog named 'Julius Caesar'. And if George adds 'Julius Caesar is a dog', the ambiguity may be deemed to be wholly resolved. Of course, if George is reasonably demented, one could still take him to be referring to the ancient Roman. In which case, his use of the present tense 'is a dog' would have to be construed as a stylistic device to lend vigor to an abusive characterization. Or conceivably, being demented, George might think Caesar is still alive, or has been reincarnated as a dog and so on.

'Hamlet is an heroic figure' I say, and then add, 'and also a small town in North Carolina': what am I referring to? The Prince of Denmark is not a small town, not the home of John Coltrane, and the town is not a prince. But considerations of coherence readily impose an order on this apparent confusion. Not Hamlet, the Prince, and not Hamlet, the town, but the name, 'Hamlet', is being referred to, no doubt somewhat obliquely, while 'is' is an ellipsis for 'is the name of'.

34. Is it possible to have a referent of which one can form no coherent conception? Yes and no.

Coherence admits of degrees of some sort. Reference is precluded by total incoherence. But one can learn to live with partial incoherence (otherwise social intercourse, as we know it, would cease). The descriptive name 'Quasar 3C 273' presumably has a referent of some sort, but it is not clear that anyone has a wholly coherent conception of a quasar. Someday we may discover that we have not been referring to what we thought we have been referring to in speaking of Quasar 3C 273.

How far can one move in the direction of incoherence? Suppose the name 'Grun' is supposed to refer to a fictional entity that is a crocodile and a battle and a flock of crows and a person who is tall but short and thin but fat and so forth. Does 'Grun' refer? If 'Grun' were supposed to refer to a real person, then 'Grun' does not refer. A real person can't be a crocodile and a battle and a flock of crows and a person who is both tall and short and thin and fat. But can a fictional entity? Perhaps Grun is not one but two or three or more fictional characters or events, all with the same name? But what if these simple resolutions were precluded? Consistency in arithmetic may be important if one is building bridges; does it matter if one is using the arithmetic to form tattoos? An exploration of the limits of incoherence is a task better left to creative writers than nice-minded philosophers.

35. That coherence, rather than existence, is the critical factor in matters of reference need not be surprising to one who has reflected on the structure of natural language.
As Zellig Harris has said: "What is special to a grammatical utterance (i.e. to a linguistic event) is not that it has meaning, expresses feelings, communicates, or calls for a relevant response – these are all common to many human activities – but that it is socially transmissable".[18] Social transmissability calls for an interaction between a speaker and a hearer: the speaker in uttering an utterance may therein manage to invoke certain conditions which serve to structure and focus the attention of the hearer. (Attention, and not intention, is what is important in the use of language, as any good behaviorist knows.)[19]

36. Consider Grunt: 'Who is Grunt? Grunt is, or was, the giraffe who lived in my garden in Nairobi'. In so saying, I have not referred to Grunt: I have identified Grunt for you. Having done that, I proceed to say: 'Grunt is, or was, a tall fellow, 19 1/2 feet high'. Now I have referred to Grunt. I have invoked various conditions: those of being (or having been) a giraffe, of being in a garden, of being in Nairobi and so forth. The conditions invoked are those indicated in the preceding act of identification. Assuming that you, the reader,

attended to and understood the utterance, 'Grunt is, or was, the giraffe who lived in my garden in Nairobi', the uttering of the name, 'Grunt', in the subsequent utterance served to structure and focus your attention in such a way that, for example, you were not astonished by what I said. Which you might well have been had the initial utterance been 'Grunt is a rabbit who lives in my garden'.

For an expression to be employed as a referring expression, it must have associated with it various conditions, which conditions are then invoked by the speaker in his use of the expression.[20] If the invoked conditions form a reasonably coherent set, it should be possible for the hearer to form a reasonably coherent conception of the referent. Thus, having now learned that Grunt is a 19 1/2 foot giraffe, you might ask 'How does he, being so tall, manage to get a drink of water?' (which, as a matter of fact, is quite a complex matter). Here the anaphoric substitute, 'he', clearly serves to refer to Grunt. But if you had asked 'How does he like it in Antarctica?', I would have been at a loss to specify a referent of 'he': adding the condition of being in Antarctica to the set of conditions already invoked would seem to yield an incoherent set and, in consequence, a failure of reference.

Finally, let me tell you a few more things about Grunt: he of course is a ruminant, has four stomachs, indulges in necking, has two horns and probably bends over his left leg to drink. And he is gentle, ready to share his acacia leaves with another. However, Grunt is a fictitious personage. So it is untrue that he lived in my garden in Nairobi. But then you knew that, so I needn't apologize for deceiving you. Apart from that triviality, Grunt is exemplary.

And now you should have a clear coherent conception of Grunt, and despite his nonexistence, you really can readily refer to him if you wish.

However, before we can return to questions of what one can know, or of knowledge, having invoked and relied on a conception of coherence, precisely what is being invoked and relied on must be made somewhat clearer.

CHAPTER IV

COHERENCE

37. BEING AN ADVOCATE of coherence, as well as an adherent, indeed a devotee, in so far as my logical acumen has been equal to the task, which is not, of course, to claim it has, and having, during the years of my advocacy, frequently been confronted both with the incredibly blank awful look of complete and utter incomprehension and with the impossibility, or what then seemed to be such, of, if not converting that blank to a knowing stare, at least scrawling a few lines across the void to lend it some semblance of structure, I find it encumbent on me at this point in time to lay bare its basic bones; for coherence, despite the enormous complexity of the issues into which it enters and despite the remarkably disparate character of the matters to which it pertains, is, in essence, a remarkably simple notion.

Not the word, of course, but a conception of coherence is what is of concern here; the word itself is reasonably transparent: 'coherence' is a nominalization of the verb 'cohere', which verb is clearly cognate with such verbs as 'adhere' and 'inhere'; plainly put, to cohere is to stick together, but since one readily speaks of "the coherence of speakers", of "discourse", of "theories", of "light" and so forth, it should be evident that the nominalization can, and indeed generally does, serve to express a transfiguration of the simple conception associated with the verb.

38. To pursue a *via negativa*, for instances of incoherence, if aptly chosen, often require no close scrutiny to disclose their true character, incoherent speakers may manifest their incoherence in various ways, to mention only a few prominent possibilities: a merely mumbling man may be readily deemed incoherent, not because of any apparent

36

defect in the character of the thought he was perhaps attempting to express, but solely owing to the character of its expression, which is to say that the incoherence may be wholly attributable to phonological features of his discourse; whereas not phonological but syntactic ailments seem to afflict such a discourse as 'Being such, which, when conceived of, does not lend itself that others will comprehend, to verbal expression, it will not do'; in contrast, neither phonological nor syntactic but semantic factors serve to unglue and collapse Eisenhower's fine phrase "I couldn't fail to disagree with you less". To pass to another end of this path, some tales about time travel frequently seem to conceal, but nonetheless suffer from, a basic type of incoherence: thus, for example, one may be led to contemplate the possibility of George, the father of Josef, entering a time machine to travel back to the time of his childhood, at which time he kills the prepubescent George; hence he either commits suicide or murder: in either case, who sired Josef?

It must be noted, and not merely in passing, that, though, on occasion, the attribution of "incoherence" is distinctly perjorative, it is not invariably so: thus all light sources in which there is no phase relationship between the waves emitted by the different atoms in the source may rightly be classed incoherent, this in contrast with the coherent light supplied by lasers. The significance of this observation is exemplified in the exemplary incoherence of such a Zen discourse as 'Why did Bodhidharma come from the west? Have a cup of tea!' (Less lovely are the Occidental correlates supplied by Tertullian's curious credo "*Credo quia absurdum*" and Kierkegaard's querulous echo "I believe that which is absurd".) But the perspicuous incoherence of Zen is a matter we shall have to turn to later.

Even this brief stroll along this *via negativa* should suffice to persuade one that, if there is a univocal conception of coherence that serves to evaluate these diverse cases, its essential character cannot be displayed in any strictly linguistic analysis: it is not characterizable in simply phonological or syntactic or semantic terms.

39. I do not know if one would readily speak of a stone wall as something coherent, or, if one did, whether one would then be speaking literally or figuratively; but if this is figure, the figure is pleonastic and the pleonasm will serve my purpose here; for, at any rate, a stone wall that stands is something that sticks together: it does so for altogether explicable reasons.

Encountering an ordinary dry stone wall, somewhere in some garden or other, one would see that it is constituted of various stones, each of which has a specific size, shape and mass; each stone stands in a specific spatio-temporal relation to every other; the entire collection of stones constituting the wall has a specific spatio-temporal position in a gravitational field, and so forth. One can, accordingly, characterize an ordinary dry stone wall as a highly structured entity in a complex field. If one were to ask 'What keeps the wall together?' a reasonably adequate answer could have at least two distinct aspects, for one could invoke two quite different sorts of factors: first, one could refer to the specific structure, or features of the structure, of the entity in question; or secondly, one could refer to the relevant laws pertaining to such structures. Given that the structure is, as it is in the case of a stone wall, a physical structure, and given that what is in question is what keeps the wall together, the relevant laws are physical laws. But what is of interest here is, not what causes something to be coherent, but, rather, what constitutes being coherent? And physical laws do not enter into an answer to this kind of question.

Consider two fields: in the first one finds a dry stone wall constituted of n number of stones; in the second field, one finds what I shall call an "unwall", to wit, n number of stones scattered at random throughout the field: What is the relevant difference between the wall and the unwall in virtue of which only the former is rightly or reasonably classed as coherent? (Of course, in posing this question, I am at once imposing a restriction on the use of the term 'unwall': to employ the characterization 'unwall' without any restriction on what may be in question would be to indulge in an absurdity; but I choose to construe "unwalls" along the lines of "unbirthdays": The number 7 is not an unbirthday and neither is it an unwall.) Incoherence is akin to chaos and poses some of the same problems: if one attempts to think of chaos as the total lack of order then, of course, there is no such thing as chaos; for in so far as there are some sort of elements in question, the elements will be ordered in some way, albeit randomly; incoherence conceived of as the total lack of coherence is as impossible as chaos similarly conceived; for, again, in so far as there is something in question that is incoherent, it will have to be sufficiently coherent at least to be that which is incoherent. Just as, rightly understood, "total chaos" is not the total lack of order, but, rather, the total lack of any relevant order, so, rightly understood, "total incoherence" is not the total lack of coherence, but, rather, the total lack of any relevant coherence: thus, although the discourse of a mumbling man may be totally incoherent, that which is in question is at least sufficiently coherent to constitute a discourse. My unwall, as I have characterized it here, does not manifest a total lack of coherence, for it is at least a particular finite collection of stones scattered at random throughout a field; hence what is in question is at least sufficiently coherent to constitute a particular collection of stones, even though it is not sufficiently coherent to constitute a wall. But the question is: why not?

Although it may seem curious, the fact of the matter is that the particular unwall in question is, as the wall is, a highly structured entity in a complex field: it is constituted of a finite number of stones, each of which has a specific size, shape and mass; each stone stands in a specific spatio-temporal relation to every other; the entire collection of stones constituting the unwall has a specific spatio-temporal position in a gravitational field. (Given that the unwall is a discontinuous entity, how specific its specific spatio-temporal position is is far from clear; but at least the unwall is located in a particular field.) Obviously, the only relevant difference between the wall and the unwall is in the arrangement of the stones: with respect to constituting a wall, the arrangement of the stones of the unwall must be characterized as incoherent. But the question still is: why?

40. Suppose a student of arithmetic is told: 'Continue the series 2 4 6 8 10 12!' and he replies '17': what is one to say? From a purely mathematical point of view, his answer cannot be faulted: no finite sequence of numbers serves to determine a series uniquely; there are infinitely many series whose initial six elements are 2 4 6 8 10 12. Nonetheless, as we all know, the sought after and, in an ordinary case, the correct reply is '14', not '17'. What makes '14', and not '17', the correct reply? Not to digress, suppose we are driving along together in my car and suddenly I say: 'Look at the odometer! It reads 19,752.6 miles.' 'So what?' would not be a surprising reply. But what if I had said: 'Look at the odometer! It reads 88,888.8 miles'? The reply 'So what?' would not come as readily to one's lips.

We humans are fanciers, connoisseurs, of coherence: I don't know why, but we are; there is no doubt about that; coherence catches our eye, fixes our attention, focuses our mind. An odometer reading of 19,752.6 is not as fascinating as a reading of 88,888.8: that's obvious, anyone can see that; but what is perhaps not so obvious is that what catches our eye is an exemplification of a principle of logic, the principle of identity: not the number, 88,888.8, but the numeral, 88,888.8, supplies a more coherent pattern to our eye than the numeral, 19,752.6, owing to its exemplification of identity. That '14' is the correct reply to 'Continue the series 2 4 6 8 10 12!' is also explicable in terms of logical structure: in this case, however, not numerals but numbers are in qestion; the series '2 4 6 8 10 12 14' is coherent in a way that the series '2 4 6 8 10 12 17' is not. Or to put the same matter in a slightly different way, the finite set constituted by the integers 2 4 6 8 10 12 17 is less coherent than the finite set constituted by the integers 2 4 6 8 10 12 14. And the evident reason is simply that the members of the latter set can be ordered in terms of the successor relation with respect to even numbers; this is evidently not the case with respect to the former set, the step from 12 to 17 being an evident anomaly.

Perhaps it should be stressed that what is important here is not what immediately seems to meet the eye but what is actually there to be seen after a careful logical scrutiny: thus the finite set of integers 1 1 2 3 5 8 13 21 34 may seem incoherent at first blush, but a more careful examination will, or can, suffice to indicate that it is a segment of the Fibonacci series; and this is simply to say that a careful examination will, again, suffice to indicate that what is presented constitutes still another exemplification of an accordance with logical structure. And to take a still more problematic example, one knows that the finite set of integers 2 3 5 7 11 13 17 19 23 29 31 is appearances to the contrary, a coherent set; but the principle of its coherence is, at one level, elusive, at another, altogether apparent; for what is true of each number is that it is divisible only by 1 or by itself, which is to say that each is a prime, which is to say that each has precisely the same property, and, furthermore, the members of the set can be ordered in terms of the successor relation with respect to primes.

41. But contrast two finite sets of integers, *alpha* and *beta*: *alpha* is constituted of 3 5 7 11 13 17 19, *beta* is constituted of 3 4 6 17 19 32 81: *alpha* is a proper subset of the set of primes, whereas *beta* is designed to be simply a random set of positive integers. Accordingly, I am inclined to characterize *alpha* as more coherent than *beta* in that *alpha* exemplifies greater accordance with logical structure. But now this is something that wants probing in furher depth. The members of *alpha* have a certain property in common, for they are all alike in being primes; but the members of *beta* also share certain properties, indeed infinitely many properties: for all members of *beta* are alike in being unequal to 2, to 20, to 200, and so forth. Being sets of integers, the members of both *alpha* and *beta* can be uniquely ordered in terms of the relation of being greater than; but unlike the members of *beta*, the members of *alpha* can be ordered in terms of the successor relation with respect to primes: it is for this reason that *alpha* can be said to exemplify a greater accordance with logical structure.
Here, however, one may be inclined to object that although the

members of *beta* cannot be ordered in terms of the successor relation with respect to primes, they can be ordered in terms of the successor relation with respect to what may be called "glugs", where glugs are positive integers belonging to the series 3 4 6 17 19 32 81 30 40 60 170 190 320 810 300 and so on. The correct reply to this objection is, I believe, that even though the members of *beta* can be so ordered, the series of glugs is less coherent than the series of primes: for, unlike the series of primes, an initial segment of the series of glugs fails to exemplify any coherent principle of construction; but, of course, owing to its circularity, such a reply fails to resolve the issue: nothing prevents one from defining a function "glug" such that $glug(n) = y$, where y is the n'th member in the series of glugs: the function glug then supplies a principle of construction: that the principle is not altogether coherent remains to be seen.

To see that it is not altogether coherent, one must take a closer look at the function glug; glug is defined by the ordered pairs: $[1, 3]$, $[2, 4]$, $[3, 6]$, and so on, such that $glug(3) = 6$ and $glug(4) = 17$. But why, if $glug(3) = 6$, does $glug(4) = 17$? Contrast this question with the analogous question with respect to, say, the series of even numbers: if the 3rd member of the series of even numbers is 6, why is the 4th 8? An obvious answer is that one generates the even numbers by determining for each successive positive integer whether it is divisible by 2; the 3rd such integer is 6 and the 4th is 8; thus one performs the same operation on each successive positive integer.

It is the conception of the same operation that is crucial and essential here; and it is this conception that occasions the difficulty with glugs. Is there any clear sense in which to determine, for any n, whether n is a glug one performs the same operation? There seems to be only one such, namely that in which one checks to see, for any n, whether n is a member of the set *beta* or of the sets constituted of multiples of 10 of the members of *beta*. But since what is at issue here is relative coherence, there is a significant difference to be noted between asking 'Is n divisible by 2?' and asking 'Is n equal to 3 or 4 or 6 or 17 or 19 or 32 or 81 or any multiple of 10 of any of these?' To ask of any n

whether it is divisible by 2 is to ask the same question in every case in a way that to ask of any *n* whether it is equal to 3 or 4 or 6 and so on is not: for no matter how one phrases it, the latter is not a single question but a series of questions, different questions joined together, but different questions nonetheless. One can conceal this diversity with an appropriate terminology: thus employing the term 'glug', one can ask, of any *n*, whether it is a glug, just as one can ask, of any *n*, whether it is an even number, but the diversity remains despite the similarity of terminology.

The relative incoherence of *beta*, and of the associated glug function, the series of glugs and so forth, is attributable to the limitations to be found in exemplifications of the logical principle of identity: coherence is a matter of logical structure, and when single entities are in question, the only aspect of logical structure that is relevant is that supplied by the principle of identity.

42. The principle of identity is, of course, $a = a$, where identity is a relation. To say that, in contrast with *alpha*, the set *beta* displays limitations in exemplifications of the logical principle of identity is not to deny that the members of the latter set are identical with themselves; it is to say, rather, that relations between successive members of *beta* are not identical with one another; the relation between, say, the first and second member of *beta* is not the same as the relation between the second and third member of *beta*.

(To speak of "relations between successive members of a set" is, of course, to speak somewhat vaguely; for example, one can hardly deny that each member of some arbitrary set that does not contain the number 73 exemplifies the same relation between each successive member of the set: for one can say that the second member of the set stands in the relation to the first of being like it in not being the number 73, and the same is true of the relation between the third and second members, and so on. But the relation of being alike in that neither is identical with the number 73 is not a relation which holds between successive members of such a set in virtue of their being successive members of the set.)

It is for this reason that the set of integers 2 3 5 7 11 13 17 19 23 29 31 may seem, at least at first glance, to be incoherent: for it is not at once evident that relevant relations between the successive members, which is to say, relations which hold between successive members in virtue of their being successive members, are at all the same. But, of course, analysis reveals that, just as 7, say, stands in the relation to 5 of being the next prime after it, so 11 stands in that relation to 7, 13 to 11, and so on.

43. To return to our unwall, the relative incoherence of the unwall is attributable to precisely the same factors as the relative incoherence of the set *beta*: the limitations to be found in exemplifications of the logical principle of identity.

Let us say that two stones are "transitively in contact" if either they contact directly or they contact stones that contact each, regardless of the number of intervening stones. Then relations between the stones of the wall, unlike those between the stones of the unwall, are alike in that each stone stands in the relation of transitive contact with each other. Furthermore, relations between the stones of the wall, unlike those between the stones of the unwall, are also alike in that each stone is in contact with some other along a vertical plane, and each is also in contact with some other along a horizontal plane (all this being predicated on the assumption that the wall in question is a well constructed dry wall, an exemplar of the stone mason's art). If our unwall were not a random collection of stones in a field, but, say, a single pile of stones, there would then be a corresponding increase in coherence. (There are nice questions here, that one could pursue, but I shall not, about why the familiar geometric solids are exemplars of coherence, and whether the sphere, for example, is the most coherent of all such entities.)

44. It is perhaps surprising that, when the coherence of single entities is in question, the only aspect of logical structure that is relevant is

that supplied by the principle of identity; but this appears to be true regardless of whether the entities are physical entities, such as stone walls, or abstract entities, such as sets of integers, or even symbolic entities, such as expressions in symbolic systems.

For consider a mumbling man whose discourse is incoherent owing to his mumbling: this is simply to say that one cannot identify the words being uttered; and this is to say that the identity of the words uttered is in question. In so far as they fail to have an identity, the discourse must be deemed incoherent. To say that they fail to have an identity is not, of course, to suggest that the sounds of the mumbling man are not identical with themselves: it is to say that they are not identical with any words of the language in question. Thus, for example, if I fail to identify the words of a discourse owing simply to my unfamiliarity with the language, I should hardly be warranted in deeming the discourse incoherent. What is required is that they genuinely fail to have an identity. In contrast, if someone says 'The man to George to wants the store go,' not the identity of the words but the identity of the syntactic structure is called into question: what is the subject or the object or the indirect object of the sentence? Again, when one hears the sentence 'I watched her duck when they were throwing rotten eggs,' one may take 'duck' as a verb; but when the discourse continues with 'It swam out to the middle of the lake,' in the interest of coherence, one reconstrues 'duck' as a noun, for otherwise one would then be at a loss to identify the expression for which 'it' is an anaphoric substitute.

Consider the two utterances 'The man will take you to the airport if ...' and 'The man said the ...': each is an incomplete sentence; but what is said in the second case is, I am inclined to think, and would, I think, ordinarily be judged to be, less coherent than what is said in the first case. If this is a correct view of the matter, the reason for its correctness is again to be found in connection with the principle of identity. Although in neither case can one actually identify the missing segment of the sentence, the possibility of what the missing segment expresses, or serves to express, is more narrowly circumscribed in the first case: a condition, of some sort, is required; whereas in the second case, although the missing segment could serve to express a condition, as in 'The man said the mortgage will be foreclosed if it is not paid today', the missing segment could instead serve simply to refer to something, as in 'The man said the name'. In consequence, there is greater uncertainty about the identity of what the completion of the second utterance might serve to express, and hence the lesser coherence of what is said in uttering that utterance.

45. The importance of the logical principle of identity in matters of coherence can hardly be exaggerated. Not only is it the fundamental principle with respect to the coherence of single entites, but it is also the fundamental principle underlying the coherence of all symbolic systems.

Years ago (in 1932) Alonzo Church pointed out that a fundamental presupposition of formal notation is that an 'a' on the left side of a page be identical with an 'a' on the right side of the page: without some such presupposition, any formal proof fails. Contrast the two "proofs": first, *p, and if q then r, therefore s*; and secondly, *p, and if p then q, therefore q*: without further specifications, the first argument must be judged incoherent as an argument, and certainly invalid. But without the presupposition that the first and second '*p*'s are identical and that the first and second '*q*'s are identical, the second argument would also be incoherent and invalid.

Appearances to the contrary, one should not be tempted to suppose that the presupposition of the identity of symbols is flaunted in natural languages: it is true that 'convict' on the left side of a page may be phonologically, syntactically and semantically distinct from 'convict' on the right side of the page; homonomy and polysemy are characteristic features of natural languages; but that the identity of symbols be made manifest at the graphological level is a desideratum only of a formal notation. (Which is not to deny that, on occasion, other considerations may prevail: thus in APL, an IBM programming language, it is possible to specify the values of variables *a* and *b* such that $a = b$, but, nonetheless, a certain function, namely the rank of the rank of a, does not equal that function of b, the reason being that APL distinguishes between scalars and vectors, but the distinction is not explicitly indicated in the notation when the vector has only one member.) The resources of natural languages are such that even the spoken utterance 'I can't bear to bear a bare bear bare' need pose no great problems of identification. And despite the homophony of the words 'Whig', as opposed to 'Tory', and 'wig', as a hairpiece, the discourse 'Is he a Whig? Yes, he's a Whig' is evidently coherent, but only on the presupposition that the first and second occurrences of 'Whig' are the same, and of course, anaphora here serves to establish that 'Whig's and not 'wigs's are in question.

That graphological identity does not, in a natural language, guarantee symbolic identity suffices to explain the incoherence of the following argument: The Morning Star and The Evening Star are one and the same planet, namely Venus; The Morning Star is a lovely sight: therefore The Evening Star is a lovely sight. The argument may seem to be of the form $a = b$, $F(a)$, *therefore* $F(b)$, but of course it is not: it is actually of the form $a = b$, $F(G(a))$, *therefore* $F(b)$ (where $G(x)$ serves to symbolize 'x is an appearance', and $F(x)$ serves to symbolize 'x is a lovely sight'): for the graphological identity of the two occurrences of 'The Morning Star' does not suffice to establish their symbolic identity. The first occurrence does not, but the second does, refer to the appearance of Venus when seen in the morning: hence the argument embodies a fallacy of equivocation.

46. As some of the preceding examples will already have indicated, sets of elements of a system, unlike single physical entities such as walls, or single abstract entities such as sets of integers, may also exemplify aspects of logical structure other than those pertaining to the principle of identity.

For example, a set of elements of a symbolic system, such as a complex discourse, may exemplify various principles of inference. Thus consider the discourse 'Are you going to play tennis? It's

snowing in Tibet': the evident incoherence of the discourse disappears if one is warranted, perhaps by background information, in
construing it as essentially enthymematic in form; perhaps the
respondent is a Tibetan Buddhist and it is known that if it is snowing
in Tibet, he does not play tennis; in which case the discourse can be
seen as exemplifying *modus ponens*, albeit in suppressed form, and so
as coherent. Again, the mild incoherence of 'George is at home. Josef
isn't. Sidney is eating figs.' is alleviated if inferential connections can
be supplied: perhaps George supplies Sidney with figs, but only when
Josef isn't at home. One could, of course, go by another route:
perhaps what is said is said in reply to 'Tell me something about
George, Josef and Sidney!'. Or, to take a more extreme case, consider
the discourse 'It is snowing in Tibet. The moon is round. Cats eat rats.
Why is 17 the arbitrary number?': without supplementary information, such a discourse is patently incoherent. But if what was said was
said in response to 'Tell me what thoughts are crossing your mind!',
the reply can be construed as exemplifying a series of inferences: if
a thought crosses my mind, I am to express the thought; the thought
that it is snowing in Tibet crosses my mind; therefore I am to express
the thought; to express the thought that it is snowing in Tibet, I say
'It is snowing in Tibet'; therefore I say 'It is snowing in Tibet'; and
so on. It should be noted, however, that even though such a construal
renders the discourse coherent, unlike the discourse, the series of
thoughts expressed must, nonetheless, in default of further information, be judged incoherent: there need be no problem in providing a
coherent account of something incoherent (and as incompetent
reviewers prove daily, incoherent accounts of coherent matters
abound).

The incoherence of a series of thoughts is of a piece with the
incoherence of theories and conceptions: they may fail to exemplify
logical principles, as in the preceding example, or they may positively
exemplify violations of logical principles. Thus the conception of time
travel, which allows George, as the father of Josef, to alter his own
causal history so as to preclude the possibility of his fathering

Josef, leads one to the contradictory conclusion that George did and did not father Josef, and thus constitutes a flagrant violation of the principle of noncontradiction: such a conception of time travel is utterly incoherent.

47. That a conception, or a theory, is incoherent does not mean that it is not possible to form such a conception, or to adopt, at least for a time, such a theory. I have heard it said that Lesnievski said that he liked inconsistent systems: they are so powerful! And an early version of Quine's *Mathematical Logic* was, after a time, shown to be inconsistent when Rosser derived the Burali-Forti paradox in it; but this is what history teaches: Russell's paradox scuttled Frege's ship, while Gödel's result dismantled Hilbert's program.

There are at least three distinct ways of dealing with incoherence or possible incoherence. The first is to exploit it, but only to a limited extent. That is the way of some Christian theologians: making a virtue of necessity, Kierkegaard rejoiced in evidence contrary to his religious beliefs, took tenacity to be a testimonial of faith; but even Kierkegaard, presumably, would have wished to contain the incoherence, would have resisted the claim that faith itself is absurd. The second way is the way of the Zen Buddhist: to exploit incoherence to the hilt: the aim, apparently, is to transcend coherence, to achieve "headlessness". In Zen dialogues there are standard questions, such as "Why did Bodidharma come from the West?", but no standard answers. For what is wanted as an answer is not a coherent answer, such as 'To bring Zen to Japan from India', nor a merely incoherent answer, such as 'Today is Tuesday': coherence is to be transcended, a perspicuous incoherence must be achieved. But this means that Zen, so conceived, can be a practice, a way of life, but never a doctrine. For even an incoherent response is at once a step in the direction of coherence, obviously so, if what is wanted is an incoherent answer. (In consequence, one might reason that a coherent answer would then constitute a deeper incoherence; but this line of reasoning starts one on an infinite regress.)

The third way of dealing with incoherence or possible incoherence is the only rational way: one attempts to eliminate it and, failing that, to live with it, contain it, in so far as possible. An example is to be found in astrophysics: current views about quasars appear to be incoherent; given the Doppler effect, one concludes that a certain quasar must be at a distance greater than, say, *gamma*, from our solar system; but given the known sources of energy, the strong interaction, weak interaction, and so forth, the energy supplied by the quasar indicates that it must be at some distance less than *gamma*. So something is wrong and astrophysicists look for ways of resolving the matter. As far as I know, they are still looking: one learns to live with problems.

48. Coherence is a matter of logical structure, but when what is in question is the coherence of logic itself, the difficult questions of coherence are transformed to questions of completeness and consistency. Given the Gödel theorem, one must conclude that no classical system of logic (one sufficiently powerful to account for classical mathematical structures) can be proven to be coherent, for the possibility of a consistency proof has been ruled out. (There are complications that, perhaps, should be mentioned here. There are some systems of logic that appear to be exempt from Gödel's proof: Fitch's *Basic Logic*, for it lacks negation; Church's logic derived from his *Lambda Calculi*, which has infinitely many quantifiers. But, intuitively speaking, such systems seem, in consequence, to suffer from vagueness, and the presence of vagueness would seem to preclude the possibility of any clear proof of coherence.)
But should one conclude that classical logic must be judged at least minimally incoherent, on the grounds of incompleteness? Here I am inclined to say that, prior to the Gödel result, the answer to this question should have been 'Yes', but, given the Gödel result, the answer must now be 'No'. For a conception of coherence, as any conception of anything, is not something immutable in time, fixed forever: it is subject to evolution and growth, to revision. Gödel's

proof is a proof that, if coherence is to be a coherent conception, perfect completeness cannot be required: the ideal of completeness must be replaced with the lesser goal of maximizing completeness.[21]

CHAPTER V

HYPOSTASIS

49. CONSIDER the discourse: 'Did you know that McEnroe beat Borg at the U.S. Open? Yes, McEnroe's beating Borg dismayed me'. 'McEnroe's beating Borg' is another nominalization of the sentence 'McEnroe beat Borg'.

In the discourse in question, instead of saying 'McEnroe's beating Borg dismayed me', one could just as well have said 'That McEnroe beat Borg dismayed me': there would have been only a stylistic difference.

If we suppose that 'that McEnroe beat Borg' refers to something, what are we to suppose it refers to? Let's suppose it refers to a "fact", an abstract entity. Are we then also to suppose that 'McEnroe's beating Borg' also refers to a fact? McEnroe's beating Borg might have caused a riot, but abstract entities do not cause riots. Hence it seems more plausible to suppose that, if 'McEnroe's beating Borg' refers to anything, it refers to an event rather than a fact. But then, what are we to say of two speakers, one of whom says 'McEnroe's beating Borg dismayed me', while the other says 'That McEnroe beat Borg dismayed me'? Is one dismayed by an event, the other by a fact? Furthermore, even though one can say both 'McEnroe's beating Borg was an accident' and 'That McEnroe beat Borg was an accident', one does not, in SNEAE at any rate, say 'That McEnroe beat Borg caused a riot' even though one can say 'McEnroe's beating Borg caused a riot'. If one supposes that the deviance of 'That McEnroe beat Borg caused a riot' is to be explained by saying that 'That McEnroe beat Borg' refers to a fact, an abstract entity, and that abstract entities cannot cause riots, then one is confronted with the question: how can an abstract entity be an accident? And if I say 'It is deplorable that McEnroe beat Borg', am I deploring a fact or the fact that McEnroe beat Borg?

50. Consider the remark 'I've known McEnroe to beat Borg, but never on clay': are we to suppose that 'McEnroe to beat Borg' also refers to a fact?

Notice that one can say both 'I expect McEnroe to beat Borg' and 'I expect that McEnroe will beat Borg' (or 'I expect McEnroe will beat Borg'). But though one can say 'I think that McEnroe will beat Borg' (or 'I think McEnroe will beat Borg'), one does not say 'I think McEnroe to beat Borg'. If we suppose that 'that McEnroe will beat Borg' refers, now, not to a fact, but to a "proposition", another abstract entity, but that 'McEnroe to beat Borg' does not refer either to a proposition or to a fact, then it would seem that what I expect when I say 'I expect McEnroe to beat Borg' is not what I expect when I say 'I expect that McEnroe will beat Borg': that is implausible.

Again, though one might say 'I want McEnroe to beat Borg', one does not say 'I want that McEnroe will beat Borg', whereas one can say 'I hope that McEnroe will beat Borg' even though one does not say 'I hope McEnroe to beat Borg'. And again, although one can say both 'I predicted that Borg would be beaten by McEnroe' and 'I hoped that Borg would be beaten by McEnroe', one can say 'I predicted Borg's being beaten by McEnroe', but one does not say 'I hoped Borg's being beaten by McEnroe'.

51. If one were silly enough, or perverse enough, one could attempt to draw some peculiar conclusions on the basis of the above sort of data.

So one might claim that one does not hope what one predicts; one does not predict what one hears; one does not hear what one wants; one does not want what one bets; one does not bet what one expects; and so forth. For, as has been indicated, one can say 'I predicted Borg's being beaten by McEnroe', but one does not say 'I hoped Borg's being beaten by McEnroe'; one can say 'I heard McEnroe beat Borg', but one does not say 'I predicted McEnroe beat Borg'; one can say 'I want Borg to beat McEnroe', but one does not say 'I hear Borg to beat McEnroe'; one can say 'I bet Borg has beaten McEnroe', but

one does not say 'I want Borg has beaten McEnroe'; one can say 'I
expect Borg to beat McEnroe', but one does not say 'I bet Borg to
beat McEnroe'; and so forth.

Before embracing these conclusions, let's look to another simpler sort
of case to gain a perspective on our curious data.

52. George is standing before me: he has become incredibly obese.
I say to him 'That is deplorable' and he replies 'What does 'that' refer
to? What is deplorable?'

I might answer his query by saying 'That you are obese', in which case it
would seem that 'that' refers to a fact. Or I could reply 'Your obesity', in
which case it would seem that a trait or property is that which is being referred to by the word 'that'. Or I could reply 'Your being obese', in which
case it would seem that a state of affairs is being referred to by the word
'that'. Or I could reply 'For you to be obese', in which case I do not know
how to characterize that which 'that' is supposed to refer to other than
by saying it is that which 'that' refers to.

Consider the following discourse: I say to George 'Your obesity is
deplorable', and he replies 'Perhaps, but I am glad that you are not
deploring the fact that I am obese;' or he replies 'Well, at least you
are not deploring my being obese'.

53. Metaphysical entities are, by and large, the products of the
rhetorical device of hypostasis. Roughly speaking, hypostatization is
the process by which speakers of a natural language conjure up a term
when a term is wanted, which is to say, when one wishes to structure
the attention of the participants in a discourse in a certain way.

One utters the sentence 'George is obese', and what one says in so
doing may be true: George stands before one, and one says, perhaps
sotto voce, or to oneself, 'George is obese'. There is, as it were, a
mountain before one, and one wishes to chart, to map, only certain
features of the terrain. There are many ways of map making: one can
employ a Mercator projection, or an Isometric projection; one can
construct an Equal Area map, and so forth. Nominalization is a

process, a lingustic process, by which one creates nouns out of nonnouns: hypostasis is the rhetorical device by which one conjures up referents for the products of nominalizations.

Map making is not world making: neither is nominalizing and hypostatizing a matter of creating entities: conjuration is not creation. (The device of hypostasis is a rhetorical device: not, as some would have it, an ontological commitment.) Which nominalization and consequent hypostatization one opts for in a particular situation is of a piece with which projection one elects to use in mapping a given terrain. Our choices are dictated by the verbs we employ and our choice of verbs is dictated by our interests.

George is obese and, say, that is something I do not want: then I can say 'I do not want George to be obese'. But if the verb that answers to my interest at the moment is the verb 'deplore', then, not 'George to be obese', but 'George's being obese' is wanted as an object for the verb, and so 'I deplore George's being obese'. But this is not to say that if deploring is what I am about, then I must eschew a use of the expression 'George to be obese': not at all, for there are ways and ways of deploring matters. So one can say 'For George to be obese is deplorable', and in so saying one may be deploring what one wishes to deplore.

'Deplore' is a fairly free verb: one can deplore almost anything. But 'wish' and 'hope' are much more restricted. 'I wish George not to be obese' is fine, but 'I hope George not to be obese' is deviant. On the other hand, the nominalization 'that George is not obese' is readily available as an object for the verb 'hope', but not for the verb 'wish': 'I wish that George is not obese' is deviant in SNEAE, or at least in my idiolect. But again, if wishing is what one is about, and if one wishes to use the nominal, 'that George is not obese', there are ways. So one says: 'I wish it to be the case that George is not obese'.

It is a gratuitous piece of metaphysical nonsense to suppose that the use of different nominalizations indicates anything more than the adoption of different forms of projection.[22]

CHAPTER VI

KNOWLEDGE

54. IT IS NOT EASY to say what it is one is saying when one says something as simple as 'I know George'. The significance of the remark may be modulated in indefinitely many ways, depending on the circumstances in which it is made. If I have already been introduced to George at a party and the hostess approaches me with the query, 'Do you know George?', I might reply 'Yes, we've just met'. On the other hand, if a friend of mine at the party says to me 'That fellow George wants to borrow my Porsche. Is he to be trusted?', I might reply 'I don't know him; we were just introduced a few moments ago'.

The sense of 'know' indicated when I say 'Yes, we've just met' is not the sense of 'know' that warrants a use of the nominalized form 'knowledge'. Whereas the sense of 'know' in which I claim not to know George is the sense which does admit of such a nominalization. In claiming not to know him, I am claiming not to have adequate knowledge about him.

What this indicates is that in Enlish, or in SNEAE at any rate, one is not warranted in speaking of "knowledge by acquaintance" and "knowledge by description" as two kinds of knowledge. One can have knowledge of persons, places, things, events, subjects and so forth: but this is not a matter of different kinds of knowledge, but simply knowledge of different kinds of things.

55. Since what I am saying explicitly contradicts the views expressed by Bertrand Russell in his *The Problems of Philosophy*, it may help to attend, if only briefly, to what he said.

According to Russell, we have "acquaintance" with "anything of which we are directly aware, without the intermediary of any process of inference or any knowledge of truths".[23] And, according to Russell, in the presence of a table, I am acquainted with the "sense-data" that make up its appearance; in particular, say, its color. He claimed that "so far as concerns knowledge of the colour itself, as opposed to knowledge of truths about it, I know the colour perfectly and completely when I see it, and no further knowledge of it itself is even theoretically possible".[24]

The difficulty I find with Russell's view is that his use of the words 'know' and 'knowledge' is, from the standpoint of SNEAE, utterly bizarre. Suppose I have been painting in oils for some time, but with a very limited palette; someone then gives me a tube of Windsor-Newton Naples Yellow, a pigment I have not ever used before. Squeezing some out and spreading it on my palette, I might exclaim, 'I know that color! I've seen it in one of Piero's works'. I would say that I know the color, or that I am acquainted with the color, if I have seen it before. But if, after spreading it out on my palette, I say 'I have never seen this before: it's a marvellous color', it would be absurd for someone hearing what I have just said to ask 'Do you know this color?' Evidently I do not know the color, though, perhaps, henceforth I shall be able to say that I do. And if, some time after, an artist friend were to squeeze out some Naples Yellow on his palette, I might say 'I know that color; isn't it superb!' Russell said that "I know the colour perfectly and completely when I see it", and I am saying that, if I have never seen the color before, I do not know it, but, perhaps, the next time I see it I may be able to claim that I know it. There is a diachronic aspect of the use both of 'know' and 'acquainted with' that Russell seems to have been completely oblivious to.

The oddity, and implausibility, of Russell's claim that "I know the colour perfectly and completely when I see it", is, perhaps, more apparent if one thinks, not of knowing colors, but of knowing pitches. Do I know Middle C, perfectly and completely, when I hear it? Unfortunately, I do not, for I do not have perfect pitch. But this is not to say, if someone sounds Middle C on a properly tuned instrument, that I do not hear the note that is sounded. I hear it, but,

even so, if, a few minutes later, one were to sound the same note
again, and I were asked, 'Is that the same note that you heard
before?', quite possibly I should have to reply 'I don't know'.

Furthermore, even if in looking at a color I am in a position to say
'I know that color', there is nothing, in such a situation, that warrants
a use of the nominalization 'knowledge'. If someone were to spread
out some Naples Yellow on my palette, I might very well say 'I know
that color; I use it all the time', but, in so saying, I am not claiming
knowledge of any sort whatever. Nor is there any reason to attribute
to me knowledge of any sort. One would be warranted in attributing
some sort of knowledge to me only if I could say something about
Naples Yellow, other than that I know the color; for example, is it
warm or cool? Is it dissonant with Cadmium Yellow Pale?

What is truly peculiar about Russell's view is that he wishes to insist
that knowing a color is not a matter of knowing any truths about the
color, but is, nonetheless, a matter of knowledge. But such "knowledge" is not simply not socially transmissable, it is theoretically
unstateable: such a conception of knowledge has, surely, little utility.
Consider how curious the situation would be if someone, alarmed by
the presence of a strange odor, thinking perhaps that it might be a
noxious gas, were to ask me, 'Do you know that smell?', and I were
to reply in a Russellian vein, 'I have perfect and complete knowledge
of the smell, but, unfortunately, I know nothing about it'.

Conceivably, all that is at issue here are differences between SNEAE
and English English at the turn of the century. But I doubt it.

56. If one knows that something is the case, then, speaking abstractly,
one has knowledge of a truth. If the truth in question is a truth of some
particular subject, then possibly one has some knowledge of that
subject. (How many truths one must know of a particular subject
before one can sensibly be said to have at least some knowledge of
the subject is not a question that admits of any clear answer. If a small
child seems to know that one plus one equals two, would one say that
the child seems to have some knowledge of arithmetic? I don't know.

But, certainly, an adult who knows that there are infinitely many primes would be said to have some knowledge of mathematics. However, if the person were to say, 'I know that there are infinitely many primes, but I must admit that I haven't the foggiest idea what a prime is', that he had any knowledge of mathematics would at once be in doubt, though one might grant that he had some slight knowledge about mathematics.)

Knowledge of a subject calls for knowing truths of that subject. But knowledge of persons does not call for knowing truths "of" the person; rather it calls for knowing truths about the person (which is not to deny that one can have, say, knowledge of his whereabouts and the like).

57. Knowledge about a person and knowing a person are, however, markedly different matters; how different they are can be seen in the fact that, though I may have considerable knowledge about George's dead uncle, I would not now say I know him.

One cannot know a person who is dead. The apparent counter-examples to this claim are hardly worth considering, but we can glance at a few.

Of course one can know Dante, but that is to know Dante's work or who he was. One can know the dead on a battlefield, but that is to know who the dead are. One can even say, 'I know my Grandfather: that is just the sort of trick he'd play on his heirs', but that is to know my Grandfather's character. Of Yorick, Hamlet rightly said, 'I knew him', not 'I know him', whereas I am inclined to say of Hamlet, 'I know him', not simply the play of that name, but Hamlet, The Prince of Denmark: but then he's not dead, even though he died in the play. After all, Hamlet is the Prince of Denmark, not 'was'. But, of course, if I know him, what I know is the character of the character in the play.

58. That, in saying 'I know George's uncle', while George's uncle is alive, I need not be claiming knowledge of any sort, is clearly

indicated by the fact that once George's uncle is dead, I might no longer claim to know him, and might, instead claim to have known him. For though one may admit that knowledge may erode in time, what knowledge I have of George's uncle cannot instantly be dissipated by his death.

If I know George's uncle, and George's uncle dies, just as death transmutes George's uncle to dust, so it transforms 'know' to 'knew'. I shall speak of such a sense of 'know' as "the acquaintance sense".

59. Knowing someone or something, in the acquaintance sense of 'know', is never a matter of having knowledge of or about anything, even though, if one does know someone or something in that sense, one is almost certain to have some knowledge about that which one knows.

If I know George only in that I have just been introduced to him, for example, at a party the hostess says to me, 'I would like you to meet George', and George and I then shake hands, then I know that the person I have just shaken hands with has been referred to as 'George'. If it should prove to be the case that what I have just shaken hands with was an animated waxwork, not a person at all, then I would not, in fact, have been introduced to anyone. Perhaps, however, it would be true to say that I know a waxwork called 'George'; but, if so, I would at least know that that which I have just contacted by hand has been called 'George'.

Or suppose I simply see a color that I know: then I am likely to know that the color I see is a color I have seen. For if I have never seen the color before, it cannot be true that I know the color. (However, I would not deny that someone might claim to know a certain color, and yet deny that he has ever seen it before; one sufficiently bemused might claim to have innate acquaintance with the color: the exploration of the fields of folly is, after all, a daily endeavour.)

60. That knowing someone or something, in the acquaintance sense of 'know', is never a matter of having knowledge of or about anything is indicated by the fact that no amount of knowledge ever warrants

the claim that I know someone or something, in the acquaintance sense of 'know'.

Suppose Josef knows, as one says, all about George. Josef knows who George is, what he does, where he lives, where he has lived, whom he has married, whom he has divorced; he has even discussed all these matters with George's psychiatrist. Furthermore, he has been at the same meetings as George, the same parties, conventions and so forth. Still more, on all these occasions he has observed George closely, heard George discuss all sorts of topics, noted George's movements: Josef is, indeed, in the process of writing a biography of George. However, George has never set eyes on Josef, has never heard of him; in short, George has not an inkling of Josef's existence. Does Josef know George?

In some sense of 'know' Josef does know George, for he could truly say to George, 'Even though you don't know me, I know you very well'. On the other hand, if Josef were to tell Sidney that he, Josef, knows George, and if Sidney were then to turn to George saying, while pointing at Josef, 'This man claims to know you', George could well respond, 'That man's a liar: I've never set eyes on him'. Does Josef know George? Though in some sense, yes, nonetheless, no, for he does not know him in the acquaintance sense of 'know'.

It is true that Josef could approach George and say 'Even though you don't know me, I know you very well'. In that sense of 'know', Josef does know George, but that sense of 'know' is a diminished sense of 'know': the preamble to Josef's remark, 'Even though you don't know me', together with the fact that the remark is addressed to George, is significant here. If Josef were to say to Sidney, 'I know George, even though he doesn't know me', Sidney would rightly take Josef to be saying that he, Josef, knows who George is, but not that Josef, in fact, knows George. Furthermore, given the preamble to Josef's remark, Sidney would take Josef to be saying, not simply that he knows who George is, but that he knows a great deal about George; even so, Sidney would not take Josef to be saying that he, Josef, in fact, knows George.

Consider the perhaps less confusing case of a celebrity, say, Rod
Laver, seated in the stands watching a tennis match. Josef and a
person alongside him are also watching the match; Josef, spotting
Laver, turns to the person next to him saying, while pointing at Laver,
'I know him'. Clearly, Josef's claim is ambiguous: he could be saying
that he knows who the person is that he is pointing at. In that case,
his neighbor might nod, saying, 'Yes, that's the Rocket'.[25] On the
other hand, he could be claiming to be acquainted with Laver. In
which case, his neighbor might say, 'Really? Where did you meet
him?' Or his neighbor might turn to Laver saying, 'Hey Rocket, this
bloke says he knows you', and Laver might reply, after taking a hard
look at Josef, 'I've never set eyes on him, Emmo'; whereupon
Emerson might turn to Josef with a grin saying, 'If you really want
to meet him, I'll introduce you after the match'.

61. That knowing, in the acquaintance sense of 'know', is not a matter
of having knowledge is, or should be by now, reasonably clear. But
it is important to realize why this is so.
One is warranted in speaking of "knowledge" only if that which is in
question is at least theoretically socially transmissable.
Suppose you and I are looking at colors spread out on an artist's
palette, and I say, pointing at one of the colors, 'I know that color',
if I were claiming knowledge of any sort, what could I be claiming?
It is true that if I know the color, then I've seen it before. But it could
be true that I've seen it before and yet not true that I know the color.
Furthermore, I might even know that I have seen the color before,
because you know I have and you have just told me so, and, even so,
I might not know the color. In telling you that I know the color, I am
indicating, not only that I have seen it before, but that I know it, in
the acquaintance sense of 'know'. You might thus acquire some
knowledge about me; but in saying 'I know that color', I am not
attempting to provide you with any knowledge about the color. The
only "knowledge" about the color you could possibly glean from this
would be the fact that "the color has been seen by me at some time

or other": the implausibility of this passive should, by now, be apparent.

62. That knowledge implicates social transmissability is further evidenced by the fact that we do not attribute knowledge to animals that do not use language.
One may say, perhaps, that a dog knows the sound of his master's voice, his footsteps, his appearance, his odor and so forth. But one does not conclude: 'The dog has considerable knowledge about his master'. Or one may say that a cat knows where its food is kept, but one does not say, 'The cat has knowledge of the location of the food'.

63. Though I have just said "one does not say" such things, I must admit that some do. (Which, of course, indicates that I am using the expression 'one doesn't say' in much the same way the English are prone to use the expression 'It isn't done' in reference to what one has just done but is frowned upon.)
Talmy Givón is not the least bit reluctant to speak of canine knowledge.[26] Thus he attributes to canines "background information regarding the terrain" and speaks of this as "*stable, common knowledge*". It would be a gross error to suppose that Givón is simply misusing the word 'knowledge'. For what is at issue here is not simply the deviant use of a word, but the conceptual scheme that underlies such a deviant usage.
Givón speaks of dogs as having "some notion" of something, of canine "speech", of dogs having "some mental representation".[27] Givón thinks that in canine communication "the subjects – agents are only *you* or *me*". Thus he apparently ascribes self-awareness to canines. Given such assumptions, it is not surprising that he has let "knowledge" go to the dogs.[28]

64. There are three aspects to the possession of knowledge that warrant attention here. These are the aspects of internalization, of memory and of systematization.

Much of anyone's knowledge is internalized. Knowledge of the grammar of one's own language is as good an example as one can find: in everyday discourse one makes remarks displaying remarkably complex syntactic structures without having to pause to remember what syntactic form to employ; this is evident when one compares speaking one's native language with speaking a foreign language in the initial stages of learning.[29]

65. One cannot identify what is remembered with what is internalized: these are quite different conceptions. I remember the neighborhood where I lived as a child, but it would make little sense to say that I have "internalized" that neighborhood.

Owing, perhaps, to the development of automata technology, many, today, are prone to speak of memory as a "store", to speak of "memory banks and registers", to speak of "accessing" one's "memory store" and so forth. I am not the least bit opposed to viewing human beings as automata, with the proviso that it be recognized that what are then in question are organisms, not mechanisms. But when organic automata are in question, there is little reason to employ the figure of a "store".

Consider the difference between searching for a certain record in a pile of records and trying to recall a name. To find the record, one goes through the records, one by one, if the records have not been organized in any way, or if one hasn't a clue to the principle of organization. But if one is trying to recall someone's name, it would be futile to attempt to recite all the names one knows. Instead, one uses devices to trigger one's memory. Meeting someone at a party who evidently knows me and whom I am sure I know, but whose name escapes me, I ask myself such things as: where have I seen him before? Does his name begin with 'a', 'b', ... ? Thus one supplies oneself with stimuli that may serve to evoke the desired response.

Rather than likening memory to a "store", or to a "pushdown-register" or the like, a more plausible figure is supplied by a thin sheet of copper displaying the vicissitudes of time: a sheet with various bumps, dents, scratches, traces and so forth. One can further suppose that the various deformations of the sheet correspond to various memories, that the sheet itself is a complex function mapping stimuli on to responses, but one subject to evolution and decay in time. Let it be clear that I am not claiming that this is a correct model of human memory: I am only saying that it is more plausible than any model in terms of registers, memory banks, stores and so forth. (A still more plausible model would be supplied by a sheet, not of copper, but of some more variable unstable substance, the deformations of which would more accurately correspond to the differences between short and long term memory, a substance that would enable one to model the effects of senescence on memory.)

66. The most important aspect of knowledge is that of systematization. This is simply owing to the fact that knowledge that has been subject to systematization is both subject to growth and maximally socially transmissable.

Internalized knowledge is obviously not easily socially transmissable.[30] But although knowledge based on memory is transmissable, present processes for such transmission are laborious and boring. Learning things by rote is a dull affair, even when supplemented by such brilliant devices as Skinner's teaching machines, programmed texts and the like.

Knowledge subjected to systematization admits of genuine growth, whereas knowledge based on memory admits merely of accretion. If I know the names of forty three cities, each of which is the capital of a state, that knowledge will not enable me to know the capital of still another state. In contrast, if one learns a few principles of logic, one is able to deduce an astonishing number of interesting and informative theorems. The difference here is the difference between the knowledge displayed by "quiz-kids", of bygone television, and the knowledge of a competent logician.

67. Though the acquisition of knowledge may be of considerable importance, it is not inevitably so: some items of knowledge may be altogether trivial. (The importance of recognizing the unimportance of some items of knowledge will appear later.)[31]

The triviality of an item of knowledge is not the same as the triviality of a claim to knowledge: to make a claim is to perform a social act. The triviality of trivial claims may have various distinct sources.

A claim to know something may be trivial in that what one claims to know may be trivial. Thus if I claim to know that I have just blinked my eyes while I was writing, the claim is utterly trivial: what difference can it make to anyone, even me? (This is not to deny that there could be situations in which such an item of knowledge would be of some significance; but this is not one of them.)

A claim to know something may, however, still be trivial even if what is claimed is not itself a trivial matter. Let p be that entropy increases in isolated systems; then even so, if Josef claims to know that p, Josef is making a trivial claim. The truth of p is well known; hence if Josef is in error, the error is not likely to be transmitted to others. For if Josef is in error, the error is not an error with respect to the truth of p: the error must be owing to the fact that Josef is not in a position to claim any knowledge with respect to the truth of p: perhaps Josef believes everything he reads, and he came across the statement about entropy in a work of science fiction.

There is, however, a third, seemingly inegalitarian, source of triviality that must not be overlooked. Let p be a statement to the effect that some highly controversial principle of astrophysics is, in fact, incorrect. Then p is clearly not a trivial matter. Nonetheless, if I claim to know that p, my claim is trivial, whereas if Einstein had claimed to know that p, his claim would not have been trivial.

68. This third source of triviality exemplifies one of the many respects in which there is a striking similarity between epistemological and moral matters.

If an underprivileged impoverished person steals a book from the university library, his act is, no doubt, reprehensible. But the seriousness of the offense would be aggrandized considerably if the act were to be performed by a university professor. For not only would the professor have committed a theft, he would have failed to fulfill the obligations of his position.

If I claim to know that a certain highly controversial principle of astrophysics is incorrect, owing to the fact that I am a philosopher, and not an astrophysicist, that I have no standing in the community of astrophysicists, thus owing to the fact that no one is likely to pay any attention to what I say about such matters, my claim to such knowledge becomes a trivial matter. If I am in error, the error is not likely to be contagious. It will not give rise to misguided research.

CHAPTER VII

KNOWING HOW

69. HAVING KNOWLEDGE is sometimes a matter of knowing something about a person, or a place, or knowing a subject and so forth, but it is also sometimes a matter of knowing how to do something. Gilbert Ryle, in *The Concept of Mind*, made much of a distinction, which, since Ryle's work, has beome a commonplace in epistemology, between what is called "knowing how" and "knowing that". I have no real quarrel with the distinction, but one must deplore the characterization of the distinction, and certainly the label it bears. The word 'how' at one time was spelled 'whow': it is a pity that it lost that spelling; such a spelling served to indicate that 'how' belongs to the group of so-called "wh- words" in English, namely, 'whether', 'who', 'which', 'why', 'when' and so forth.

There may seem to be no epistemological distinction between statements of the form 'I know that ...' and of the form 'I know wh- ...'. For if I know that George did something, then I know whether or not he did it, and if I know whether or not he did it, then either I know that he did it or I know that he did not do it. But inferences do not proceed as smoothly when one shifts to other wh-words.

If Josef knows that George found a cypripedium, but Josef doesn't know what a cypripedium is, does Josef know what George found? If Josef knows that George is in the Piazza Bologna, but Josef doesn't know where the Piazza is, does Josef know where George is? (If Josef knows that George is in the solar system, does he know where George is? There is a perspective from which one could say 'Josef knows that George is on Earth', but it is not one that many of us are likely to attain in the near future.) If Josef knows that a certain battle was fought during the reign of Claudius, but he doesn't know when Claudius reigned, does he know when the battle was fought? And if Josef knows that George used a "forcing" procedure to prove the theorem, but he doesn't know what a forcing procedure is, does he know how George proved the theorem? In none of these cases can one give an unqualified affirmative answer to the question posed. Conversely, if, in the cases cited, one knows how, or when, or where or what, it doesn't follow that one has any specific knowledge that something is the case. Thus, for example, if I know where George is, and George is in the Piazza Bologna, it certainly does not follow that I know that George is in the Piazza Bologna. I may know where George is, and I may demonstrate that I do by leading you to him; that does not mean that I know that where he is at is the Piazza Bologna. Perhaps, however, one could say that if I know where George is, then I know that George is in a certain place: by retreating into vagueness, one can, perhaps, collapse the distinction between knowing where something is and knowing that something is the case. Strictly speaking, however, one can gloss 'I know where George is' as 'I know that George is in a certain place' only if the latter sentence is taken in a certain way. For even if I don't know where George is, if I know he exists, then I know that he is in some place or other. From a logical point of view, it would be more plausible to gloss 'I know where George is' as 'I know that a certain place is the place where George is'.

(Although this is something that will have to become clearer as we proceed, for now is not the moment to go into details, an attempt to gloss 'knowing where' in terms of 'knowing that' is of a piece with an attempt to adopt a synchronic view of matters that are best viewed diachronically. That I know where George is is something that, though it may resist explicit statement, here and now, may, nonetheless, be manifested in behavior, and thus in time.)

Just so, one may attempt to bridge the gap between knowing how and knowing that (given that one adopts a synchronic stance) by glossing 'knowing how one builds a bridge' in terms of 'knowing that one builds a bridge in a certain way, or in a certain manner or by certain means'. To find a less tractable distinction between expressions of the form 'know that' and expressions of the form 'know how', one must attend to intransitive verbs. But even then one can persist: 'swim' is an intransitive verb, but knowing how to swim in tomorrow's race may be a matter of knowing that one should pace oneself carefully in the first ten laps, and then go all out for the remainder of the race. There is, however, a completely intractable distinction to be found between knowing that something is the case and knowing how to do something, when knowing how is not a matter of knowing a manner or means of doing something. If, in reply to the query, 'Do you know how to swim? If so, you can join us on the canoe trip', I reply 'Yes, I know how to swim', I am not claiming that I know that something is the case.

70. But can't one construe 'I know how to swim' as 'I know that a certain way (here a certain means, a certain way of acting) is the way by which to swim'?
Suppose George claims to know how to swim: he can tell you, and in great detail, exactly how to do the crawl. But he can't and never could do it; in fact, he can't and never could swim at all. Then he doesn't know how to swim.
Does this prove that knowing how to swim is not a matter of knowing that a certain way is the way by which to swim? Just as the proof of the pudding is in the eating, so, one might argue, the proof that George does know that a certain way is the way to swim is to be found in his own performance. The fact that he can't and never could swim might seem to be a proof that he doesn't know that a certain way is the way to swim.
One might so argue, but it would be a mistake. That George can't and never could swim does not show that he does not know that a certain way is the way to swim: what it does show is that, even though he may know how one swims, he does not know how to swim. Compare: 'I know how one swims – How?' with 'I know how to swim – How?': one might ask the first 'How?', but not the second.

71. If I know how to do something, such as swim, play the piano and

the like, it does not follow that I can do it. Nor does it follow that
if I can do something that I know how to do it. For that matter, it
doesn't even follow from the fact that I did something that I can do
it: it's all too easy to oversimplify these matters.

Fooling around with a safe, suddenly I find that I have unlocked it.
Let's say that the odds against hitting on the right combination by
chance are a million to one. Later, if I am asked whether I can open
that safe, I reply 'No, I cannot'. It is not true that I can open that safe,
even though it is true that, once, by chance, I did open it.

If I am asked whether I can drink milk, I reply 'No, I cannot', for I
have lactose malabsorption. Someone else might say 'I can drink the
milk, even if he can't'. It does not follow that he knows how to drink
the milk: having lactose malabsorption is not a matter of not knowing
how to drink the milk; it is a matter of not having the appropriate
enzymes.

Imagine a scene in which a great pianist is seated at a piano, about
to begin his performance. A madman, with a Samurai sword,
suddenly appears and chops off the pianist's hands: can the pianist
play the piano? No. Does he know how? Yes.

72. 'Do you know how to breathe?': the question can seem senseless.
But whether it is senseless depends on the context in which it is
raised. If I am talking to a would-be gymnast having difficulty in doing
a series of hip swing overs on the high bar, the question is perfectly
sensible. In performing gymnastic stunts, in pressing weights, in
hitting a tennis ball, it is important to know how to breathe, to know
when to inhale and when to exhale.

Do human babies know how to suck? No, they do it spontaneously:
they are genetically programmed to do so. It is not something they
learn to do. One knows how to do something only if that which is in
question is characteristically something that one learns to do. So I
would not say 'I know how to digest the food that I eat', because I
just do it: I have never learned to do it; neither is it something that,
characteristically, one learns to do, at least, not in my culture.

But consider the cowbird: does it know how to sing like a cowbird? No, for it is genetically determined to do so. Cowbirds are parasites of a sort; they lay their eggs in other bird's nests. If young cowbirds were not genetically determined to sing like cowbirds, they would learn to sing like their foster parents. But then it would be a problem for them to meet and mate with other cowbirds. In contrast, young crows are not brought up by foster parents, and young crows learn to sound like crows by listening to their parents; indeed, there is even reason to believe that they are taught to do so by their parents.

Assume that the facts of the matter are as I have indicated. Should we conclude: 'Crows know how to caw, but cowbirds don't know how to sing'? Certainly not. But why not?

73. Even though it does not follow from the fact that I know how to do something that I can do it, witness the handless pianist, it does follow that if I do not know how to do something then I do not make a practice of doing it.

What about a person who says 'I don't know how to play tennis, but I play every day'? If he really doesn't know how to play tennis then he doesn't play every day: at best, what he does is try to play. If in fact he does play, and not merely try to play, then he does know how to play tennis, whether he knows it or not.

To return to our cowbird, since cowbirds do sing cowbird songs and sound cowbird calls, one cannot sensibly say 'Cowbirds do not know how to sing'. But since cowbirds do not learn to sing their songs or to sound their calls, one cannot sensibly say 'Cowbirds know how to sing'. Just so, one cannot sensibly say of human babies either that they know how to suck or that they do not know how to suck: cowbirds sing and babies suck; that's all there is to it.

74. It should not be supposed that the preceding remarks constitute a flagrant violation of logical principles or that the conception of knowing how to do something being delineated here is a logical confusion.

If someone points to a stone and says 'Either the stone is hungry or the stone is not hungry', one can reply that the speaker is either confused or failing to employ the correct nominalizations. Since stones are not the sort of things that can be characterized either as hungry or as not hungry, the truth the speaker is, perhaps, trying to express can be expressed by saying 'Either it is the case that the stone is hungry or it is not the case that the stone is hungry'. And in that there need be nothing to quarrel with.

It would be a logical confusion to suppose that 'It is not the case that babies know how to suck' is logically equivalent to 'Babies do not know how to suck'. (It has been suggested to me, by Jay Rosenberg, that I am, in fact, here rejecting the law of excluded middle; for on my account '*p or not-p*' does not always express a truth. What is not being rejected, however, is the law of bivalence: *It is the case that p or it is not the case that p.*)

75. If, in reply to the query, 'Do you know how to swim? If so, you can come sailing with us', I reply 'Yes, I know how to swim', in so replying I am not claiming to know that anything is the case. To claim to know how to swim in such a case is not to make a claim to knowledge of any sort: to claim to be a great swimmer is not, *ipso facto*, to claim to be a person of some knowledge. Even so, since knowing how to swim characteristically presupposes having learned to swim, doesn't learning to swim involve the acquisition of knowledge of some sort?

If one focuses one's attention on nonhuman animals, one is more inclined, I think, to say that learning may proceed without the acquisition of knowledge. My cat, Sam, learned to open a door by turning the door handle. Sam and her mate, Ben-Cat, learned to open the refrigerator by one pulling with her paw, the other pushing with his head, in the crevice between the door and the rubber seal on the refrigerator. I am not inclined to say that Sam knew that the way to open the door was by turning the door handle. But, certainly, Sam knew how to open the door; and certainly, and unfortunately, Sam and Ben-cat knew how to open the refrigerator.

If a young child were to do what Sam did, one would, however, be strongly inclined to say that, not only did the child know how to open the door, but that the child knew that the way to open the door was by turning the door handle. More specifically, when primates are in question, primates at the level of the chimpanzee or above, I am inclined to think that learning to do something implicates learning that a way to do something is to do what one does in doing it. I have heard a good tennis player say: 'I know how to serve, but I don't know what I do when I do it. However, I do know that a way to serve is to do what I do when I do it. So watch me serve and try to do what I do!'

76. It is not mere primate chauvinism that inclines one to say that learning to do something implicates learning that something is the case when humans are in question, but does not when cats are in question. For unlike cats, humans may be aware in doing something that it is they who are doing it.

Let me hasten to add that the suggestion, that cats are not aware in doing something that it is they who are doing it, is neither mystical nor derogatory. That cats have no self-awareness is indicated by the fact that cats do not recognize their own image in a mirror, and this despite the fact that they may use the mirror to track objects. It is also indicated by the fact that, if one inadvertently steps on a cat's tail, rather than looking to see what has happened to its tail, it is more apt simply to take off. That some primates are, but cats are not, self-aware is undoubtedly attributable to a difference in neuro-physiological structures. But there are virtues to be found in a lack of such awareness: cats are, in consequence, not susceptible to embarassment, vanity, shame or guilt.

Learning to do something, such as to play tennis, or to play a viola, characteristically requires attending to what one is doing, being aware of what one is doing, and, if need be, of correcting or modifying or altering it in some way. How self-conscious the process is no doubt varies considerably. But one would be hard put to explain such a simple matter as the fact that a novice at tennis, if he has any capacity at all to learn the game, gradually begins to control his wrist as he learns to stroke the ball more successfully, without supposing that, at some level of awareness, he has become aware of the fact that a limp wrist leads to loss of control.

Or should one say that learning to hit a tennis ball might be a matter of simple operant conditioning, without the intervention of any factors pertaining to self-awareness? Rats eventually learn to run a T-maze: one is not inclined to postulate self-awareness on the part of the rat. Does learning to hit a tennis ball require simply such simple conditioning? There's no reason to think it. Rats do learn to run a T-maze, but rats cannot learn to play tennis: one should not ignore the level of accomplishment at issue.

Perhaps tennis is not a good example here: lacking hands, there's no way a rat could manage the game. So imagine, if you will, a rat version of croquet: instead of using mallets, the rats are to use their noses to push the balls through the hoops. "Nose-croquet" would then definitely be within the limits of a rat's behavioral repertoire. Could a rat learn to play "nose-croquet"? I see no reason to believe it, and every reason to doubt it. Rats don't play games; even cats don't play games, which is not to deny that cats play.

Why don't rats play games? Because, pace Wittgenstein, games involve coping with artificial constraints, and for a rat, no constraint is ever an artificial constraint: an artificial constraint is akin to a self-imposed constraint, but rats, not being capable of self-awareness, are not capable of being subject to such constraints. Cats don't play games, but they come closer to playing games than rats do; for a cat will push a ball of twine behind a door and then attempt to pounce on it, as though it were a mouse in hiding. Pushing a ball of twine behind a door is close to, but not quite the same as, the self-imposition of an artificial constraint. For the character of the constraint is not the least bit artificial: the cat's play has more the form of a rehearsal of real-life contingencies, than the character of a game.

77. What is curious about the process of learning to do something, such as play tennis, play a viola, drive a car and so forth, is that once one has learned to do whatever is in question, one often is, and perhaps often should be, totally unselfconscious when one does it. One doesn't want to think about hitting a ball when one is hitting a ball in the course of a tennis match: there isn't time to think about such matters; one wants the stroke to be as smooth and spontaneous as possible.
Driving a sports car with a stick shift, the novice may have to think about when to shift gears, when to accelerate and so forth: the expert lets his feet, hands, ears and eyes make all the decisions for him.

78. One way of characterizing all this is to say that the knowledge one acquires in learning to do something, if one is successful, is likely to be internalized. Thus if one knows how to swim, or to play tennis, or to play a viola and so forth, one has acquired considerable knowledge, but that knowledge has been internalized. In consequence, one may not be aware of, or capable of saying, what it is that one knows.
Anyone who has learned a foreign language successfully is likely to have experienced such a process of internalization. When I first began to attempt to speak Italian, I would, quite self-consciously, think in English, and then try to find an appropriate sentence in Italian to express my thought. But, after a time, a considerable time, this process ceased: I began to think in Italian.[32]
Sometimes when knowledge has been internalized, it can be made to resurface. If someone asks me to tell him the combination to my locker in the gymnasium, I may not be able to do it. But if I go to the locker, without thinking about what I am doing, my fingers will open the lock for me. And of course, then, I am able to say what the combination is. If, however, someone asks me how I know, when playing the viola, whether or not I have shifted properly from first to

third position, there is nothing much to say. I know how to do it. When I do it properly, it feels right. There are various tests that one can make, and that novices have to make in learning how to shift, but once one knows how to do it, one knows when one has done it without making such tests.

79. There are important and interesting epistemological differences between knowing that something is the case and knowing how to do something. But it is not that one is, while the other is not, a matter of knowledge. No knowledge is involved in a cat's knowing how to open a door: but a fantastic amount of knowledge is involved in knowing how to play tennis as Bjorn Borg knows how to play tennis. Tournament tennis is a mental terror: Borg's ability to concentrate simply staggers the imagination. (The fairly common academic assumption that professional atheletes are persons with strong backs and weak minds is a piece of consummate stupidity.)
The first important difference between knowing that something is the case and knowing how to do something is that, in the latter case, the knowledge involved is likely to be internalized. To put the matter somewhat figuratvely, the knowledge involved in knowing how to do something is likely to be in one's fingers, the knowledge involved in knowing that something is the case is likely to be in one's tongue (but not necessarily on the tip of one's tongue).
It should be noted, however, that these matters can be somewhat reversed: knowing how to open a combination lock is essentially knowing that a certain sequence of movements will suffice to open the lock: it may be internalized nonetheless. And when one attends to the knowledge possessed by a great tennis coach, such as J. Leighton, or by a great violin instructor, such as I. Galamian, one realizes that they manage, magically as it were, to externalize and render accessible knowledge that is largely inaccessible to ordinary persons. But this calls for very special talents: one should not suppose that just because someone is an expert practioner, he is, therefore, able to teach what he practices. The fallacy in such a

supposition is more than adequately testified to by the not infrequent inanity of a painter's discourse about his own work.

Because the knowledge involved in knowing how to do something is largely internalized, in one's fingers but not in one's tongue, it is not readily socially transmissable. Some of it can be gleaned by observing the performances of professional athletes, musicians, artists, masons and so forth. Some of it is encoded in products produced by such persons: in paintings, scores, records, walls and the like. But for the most part, the knowledge resides in the fingers of the performers, and is lost when they disappear.

The second important difference between knowing that something is the case and knowing how to do something is that in claiming to know that something is the case, one is claiming to have knowledge, whereas in claiming to know how to do something, one is not. That this is so should not be the least bit surprising; for in so far as the knowledge involved in knowing how to do something is not, owing to its internalization, likely to be readily socially transmissable, there would be little point in construing a claim to know how to do something as a claim to knowledge. Claims to knowledge are of interest only when the knowledge claimed can readily be imparted to others.

CHAPTER VIII

VARIOUS USES

80. SAYING that one knows that something is the case need not be confused with knowing that something is the case. (To avoid excessive repetitions of the sentence nominal, 'that something is the case', I shall also speak of knowing that p, or knowing p, where the letter 'p' is a dummy for some appropriately formed sentence or sentence nominal, and including expressions customarily incorporated into SNEAE, such as 'Jan. 23rd is Hilbert's birthday', '$2 + 2 = 4$', and so forth. (There is no reason to confuse, and good reason not to confuse, '$2 + 2 = 4$' with 'Two plus two equals four'.)[33] Furthermore, again in the interest of brevity, I shall also speak of p's being true, or false, or in doubt and so forth, even though I would not suggest that sentences are the bearers of truth, the subject of doubt and so forth. A remark to the effect that p is true is to be taken merely as a succinct way of saying that what is said in uttering p under appropriate conditions is true.)

There are all sorts of things that I know, but which it would, ordinarily, be perfectly silly to claim to know: I know that I am alive, that I will die, that I exist and so forth. But if one is engaged in epistemic analysis, various doubts are likely to surface that do not surface when one is engaged in nonphilosophic pursuits. The only time I have ever had occasion to claim that I know that $2 + 2 = 4$ has been in the course of a lecture on epistemology or the philosophy of language. It is not something that one ordinarily claims to know. But the fact of the matter is that I do know it, despite the fact that, except when engaged in philosophic discourse, I have no genuine occasion to claim to know it.

To put the matter somewhat more technically, I am inclined to suppose that a utterance of the form 'I know that p' is likely to be semantically deviant in SNEAE when p is a dummy for such an expression as '2 + 2 = 4' and 'I know that p' is uttered by way of making a claim to knowledge. This is so despite the fact that what is said in saying 'I know that p' may, in such a case, undoubtedly be true. Indeed, it is only in those cases in which what is said is undoubtedly true that the utterance 'I know that p' is likely to be deviant: a four year old child can sensibly claim to know that 2 + 2 = 4, but, ordinarily, I cannot. That I cannot, ordinarily, sensibly claim to know that 2 + 2 = 4, however, in no way implies that I cannot claim to know that when engaged in a philosophic discussion. It would be an egregious error to suppose that philosophic discourses are, somehow, not legitimate discourses, or that, in the course of such discussions, one does not have genuine reason to make claims that one would not ordinarily make. Obviously, in discussing problems of epistemology, one claims to know that 2 + 2 = 4 in order to make the skeptical admit that there are some claims to knowledge that it is well nigh impossible to challenge.

81. Claiming to know is, in various ways, somewhat like making a promise.[34] But one need not ignore the fact that many uses of 'know' have nothing whatever to do with making a claim.

The word 'know' has a very free use in SNEAE. In the New York area, discourses such as 'I want, you know, to borrow some money, you know; I need it, you know, to pay the rent, you know; so how about it, you know?' are all too common. Apparently 'you know' functions as a parenthetic quasi-hesitation form, designed to elicit a positive response signal from the person being addressed.[35] Such discourses frequently occur in contexts in which it is perfectly clear that the hearer does not know what is indicated; for example, the hearer may know that the money is not needed to pay the rent but to pay off a gambling debt, and, what is more, the speaker may know that the hearer knows that.

The use of 'you know' as such a parenthetic quasi-hesitation form does not, in itself, occasion any problem in epistemic analysis; but

what can cause problems is the lack of a clear demarcation between
parenthetics and genuine attributions. For example, George says
'Look Josef: I'll have to charge you the full price; business is business,
you know, and I have to make a living'. Josef replies 'Yes, I know,
but I'm only asking for a 2% discount'. Is Josef here admitting that
he knows that business is business? Of course, 'business is business'
is here to be contrued along the lines of 'in business one is concerned
to make a profit and friendship is not a consideration'. Is Josef
genuinely admitting that he knows that in business one is concerned
to make a profit and friendship is not a consideration? I think it would
be a mistake to think so.

Josef's response, 'Yes, I know, but ...', hardly constitutes an
admission of knowledge; such a parenthetic 'I know' merely indicates
that the respondent is prepared to accept, at least for the moment,
the speaker's statement. This is, indeed, a frequent and common use
of the word 'know'. An unpleasant person says to me 'It would be
nice, you know, if we had lunch together today'. If I already have a
prior engagement, I then have the option of replying rudely with 'It
wouldn't be nice, and, anyway, I have a prior engagement', or, more
politely, 'Yes, I know, but I have a prior engagement'. (I don't think
one can say that the response, 'Yes, I know, but ...', in such a case
is simply a lie: on the contrary, it is much of a piece with the response
'Fine, and how are you?' said in response to 'How are you?' when,
in fact, one is anything but fine.)

82. Epistemologically interesting uses of 'know' occur not only in
connection with claims but with admissions and reports. An admissive use of 'know' occurs in such a remark as 'I knew that I had turned
off the gas, but I went back anyway to make sure'; whereas a reportive
use of 'know' occurs in 'He knew that he had turned off the gas'. In
contrast, a claimative use of 'know' occurs in such a remark as 'I
know that I turned off the gas: there's no need to go back and check'.
Fixing one's attention exclusively on claimative uses of 'know' could

lead one mistakenly to suppose that, in saying 'I know that p', the speaker is indicating a firm belief that p, or that the speaker is confident that p, or is sure that p. But the fact of the matter is that one frequently knows that something is the case, and will even admit to knowing that it is the case, and yet has no firm belief that it is so. One can even go so far as to say 'I know that she did it, for I saw it with my own eyes, but I just can't believe it. No, I don't believe it: I refuse to believe it!'

Should one say that, in such a case, the person does not in fact know what he apparently is admitting to know? Such a view would fail to do justice to the person's psychological state: the conflict he is expressing arises precisely because he does know what he admits to knowing, but, nonetheless, cannot accept. There is nothing linguistically odd about the reportive use of 'know' in 'Of course he knows that she did it, but he can't bring himself to believe it: he can't face the fact, and the knowledge is driving him mad'. Life being what it is, bounded by birth and death, occupied by stress, senescent in the grip of entropy, one knows all sorts of unwelcome truths, truths that one refuses to give credence to, to countenance or to consider.

Admissive uses of 'know' have, unfortunately, been laregely ignored by epistemologists.[36] But such uses are both frequent and useful, even necessary. If George will be forced to wait in the rain because Josef wants to reenter his house to make sure that the gas has been turned off, George may justly complain 'Look Josef, you know that you turned off the gas, so let's not waste time'. If Josef were to concede 'Yes, I know, but I just have to make sure', one then has two alternatives: either George has no warrant for his complaint, for Josef does not really know that the gas is turned off, or George is warranted in his complaint, Josef does know that he turned off the gas, but Josef is an obsessive neurotic about such matters. If Josef does not know that he turned off the gas, his behavior does not indicate that he is somewhat neurotic; whereas, if he does truly know that he turned off the gas, but, even so, feels unsure, that supports the hypothesis that he is somewhat neurotic about these matters. That is, surely, perfectly

possible. Epistemologists have no warrant to define such neuroses
out of existence.

Both admissive and reportive uses of 'know' indicate that knowing
that something is the case has very little, if anything, to do with
believing that something is the case, or being sure that something is
the case, or being confident that something is the case, or having no
doubt that something is the case, and so forth.[37]

83. Claimative uses of 'know' are similar, in various ways, to commit-
tive uses of 'promise'. Both claiming to know and making a promise
expose one to criticism: epistemic criticism in the one case, moral
criticism in the other.

If someone asks me 'Is George at home?', I may reply 'I believe he
is'. Should it prove to be the case that George is not at home, one
could say that I was wrong, that what I believed was not true; but
one could not, therewith, charge me with not believing what I had
professed. If, however, the discourse had proceeded as follows: 'Is
George at home?' 'Yes, I believe he is'. 'Listen: this is important. I
have to know. Do you know whether or not he is at home?' 'Yes, I
know that he is at home', then, should it prove to be the case that
George is not at home, I could be charged with not knowing what I
claimed to know.

The same sort of contrast is to be found in connection with making
a promise. If someone asks me 'Will you attend the meeting?', I may
reply 'I believe I'll be there'. Should I fail to attend the meeting, one
could say that I was wrong, that what I had predicted had not come
about, but I could hardly be charged with failing to keep my word.
Whereas if I had said 'I promise to attend the meeting' and then failed
to show, I could be charged with breaking my promise, failing to keep
my word, and so forth.

84. Persons often attempt to elicit claims to know in much the same
way, and for much the same reasons, that they attempt to elicit
promises. Increased assurance is what they seek.

Presumably, there is a greater likelihood that p if one claims to know that p than if one merely professes to believe that p. Just so, there is, presumably, a greater likelihood that one will perform a certain act if one promises to perform the act than if one merely predicts that one will perform the act.

In cases in which no greater assurance can be given, both claims to know and promises are idle. If, in a honest race, a jockey were to say to the horse's owner 'I promise to win the race', at best his words could only be taken as an ellipsis for 'I promise to do my best to win the race'. Were he to say 'Not only do I promise to do my best to win the race, but I promise to win it', he would simply be talking nonsense. Analogously, suppose one is about to draw a ball at random from an urn that contains 39,000 balls, numbered consecutively from 1 to 39,000; further suppose it is known to all involved that the urn contains 39,000 balls, and that one is about to make a random drawing. To claim 'I know that I shall not draw ball number 17' would, in such a situation, be pointless. If one were, in fact, then to draw ball number 17, a highly unlikely event would have occurred. But if it was in fact a genuinely random drawing, all one could possibly have known was that there was only 1 chance in 39,000 of drawing ball number 17. In such a situation, to have said 'And what's more, I know that I shall not draw ball number 17' would have added nothing at all.

85. That there are various similarities between claiming to know and making a promise should not obscure the fact that there are also important differences.

If I claim to know that p, but p proves to be untrue, one could say to me 'You did not know that p, you only thought you knew that p'. Furthermore, I might admit the error by saying 'I thought that p, but I was wrong'. Whereas, if I make a promise to perform a certain act, but then fail to perform my promise, one would not say to me, 'You did not make a promise, you only thought you did'. And even if I were to admit that I failed to perform the act in question, I would not say 'I thought I'd do it, but I was wrong'.

This difference between 'know' and 'promise' is worth attending to, for it serves to highlight the fact that knowing stands in quite a different relation to claiming to know than promising stands in to making a promise. In making a promise to do something, I am promising to do it. But, although in claiming to know that p, I am, indeed, claiming something, I am not "knowing that p".

86. There is, furthermore, an absolutely fundamental difference between claiming to know and making a promise that, for the time being, I can only point to, but not explain.

If I promise to meet someone to play tennis, but, owing to the intervention of a dire emergency, a matter of life and death, I fail to perform my promise, my excuse for failing to perform may completely exonerate me from any blame. Let us suppose that I was suddenly required to transport someone to a hospital at the time I had promised to be on court to play tennis; further suppose there had not been the slightest reason to expect any such occurrence: then not only did I have a legitmate excuse for my failure to perform my promise, an excuse that would completely exonerate me from any blame, but this failure to perform the promise in no way cast a shadow on my moral practices; it in no way implied that the assurance given by my making a promise was not what it ought to be. If I were again to make a promise to meet the same person to play tennis, he would have no reason to cast a skeptical eye in my direction when I said 'I promise to meet you'.

But the situation is subtly different when, not a promise, but a claim to know is in question. Suppose I claim mistakenly to know that George was in Chicago last week because I was in Chicago then, and I saw a person, whom I took to be George, sitting at a bar. However, unknown to anyone, even to George himself, George has an absolutely identical twin, Giorgio by name, a twin who, moreover, dresses and behaves exactly like George. The person I in fact saw was, not George, but Giorgio. George was in Yucca Flats. Then even though my mistake may be perfectly understandable, and even though almost anyone, even George's wife, would have made the same mistake, the fact of the matter is that the next time I claim to know that George was in Chicago, I am apt to be greeted, and rightly, with a skeptical look. (Precisely why this is so, however, is a matter I shall have to discuss later.)

A mistaken claim to knowledge undermines one's epistemic position in a way that a failure to perform a promise does not undermine one's moral position.

87. From this standpoint, it is not difficult to see that the definition of knowledge put forth by A. J. Ayer, in his book, *The Problem of Knowledge*, is untenable. Ayer maintained that:

the necessary and sufficient conditions for knowing that something is the case are first that what one is said to know be true, secondly, that one be sure of it, and thirdly that one should have earned the right to be sure.[38]

But consider the following case: as before, George has an identical twin, Giorgio. Both George and Giorgio are in a bar in Chicago. George is seated at one end of the bar, Giorgio at the other. Neither has seen (or ever does see) the other. Neither knows of the other's existence, nor does anyone know about both of them, except me. But I am simply someone sitting at the middle of the bar; I have been observing both George and Giorgio, and I have concluded, rightly, that they must be identical twins. However, I know neither the one nor the other, nor do I speak to either. Josef, entering the bar, sees one of them and, at once, leaves to return to Indiana; arriving there he claims to know that George was in Chicago that evening: he thus claims to know that p, p is true, and he is sure of it. So the question is: does he have the right, or has he "earned the right", to be sure of it?

Since no one (except me) knows that George has an identical twin, Josef could defend his claim to know, and claim "the right to be sure", by saying 'I saw him with my own eyes'. Given that the existence of the twin is unknown to all concerned parties, let us say the citizens of Terre Haute, George's present place of residence, Josef certainly seems to have the right to claim, as he does claim, 'I saw him with my own eyes', and this seems so regardless of whether he actually saw George or, instead, saw Giorgio. And, in consequence, he certainly seems to have the right to be sure that George was in Chicago that evening. But, even so, in such a case, it should be fairly obvious that Josef doesn't know what he claims to know.

For consider the various changes that could be rung here: first, suppose that Josef saw Giorgio, and that George was not in Chicago; instead, he was in Yucca Flats. Then even Ayer would have to say that Josef didn't know what he claimed to know since the first condition, that p be true, would have been faulted. Secondly, suppose George was scheduled to fly out of O'Hare airport an hour before Josef arrived at the bar, but there had been a flight delay and, so, George was passing the time at the bar. Then if Josef had seen Giorgio, but not George, his claim that George was in Chicago happened to be true simply because of a flight delay that he knew nothing about. Hence, it was a mere accident that Josef's claim that George was in Chicago that evening was true. Thirdly, and perhaps most importantly, suppose Josef had actually seen George, and not Giorgio; then, surely, given that he was totally unaware of the existence of Giorgio, he had every right to claim that he saw George, for only I know that it was just an accident that he was right in claiming to have seen George.

Josef's moral position was impeccable: he had every right to claim what he claimed. But Josef's epistemic position, through no fault of his own, was not adequate to warrant a claim to knowledge. Josef was not aware of the fact that he was not in a position to claim to know that George was in Chicago. That he happened to be right was a mere accident: mere accidents are not the stuff that knowledge is made of.

CHAPTER IX

CONDITIONS

88. IF A PERSON claims to know that *p*, he may be faced with the query 'How do you know that?'

Generally, one asks questions of the form 'How do you know? and 'Why do you believe?', but not 'Why do you know?' or 'How do you believe?' But this is not to deny that, on occasion, such questions are asked. Having remarked to a colleague 'It is an interesting fact that the blood pressure of the giraffe, at the level of the heart, is 260 over 160', I have met with the response 'Why do you know that?' Perhaps such a response is best construed simply as an ellipsis for 'Why do you bother to know that?'

The question, 'How do you believe that?', with reference to my remark about the blood pressure of the giraffe, would certainly be somewhat odd, but perhaps it would admit of the response, 'I firmly believe it: it was reported by a reliable physiologist'. Be that as it may, there is no doubt that interrogatives of the form 'How do you believe it?' are in fact employed. They are to be heard just about any Sunday morning in America, on the radio or on television; inevitably some ministerial type will be saying 'You believe that you are a Christian, but how do you believe it? Do you believe it with all your heart and soul?' and so forth. And then, of course, when one thinks of such types, one may well be inclined to ask 'How do they believe all that?' by way of wondering about their gullibility.

89. If one thinks about it in the wrong way, it can easily seem as if there is something odd about the question 'How do you know that?' Certainly, it is the question that one is prone to ask if someone claims to know that *p*. But even so, the question can seem to be the wrong question.

Suppose I tell someone that a capsule I am holding in my hand contains Dalmane. If he asks 'Do you know that?', I might reply, 'Yes, I do'. If he then asks 'How do you know that?', I might reply, 'The pharmacist just gave it to me and he said it was Dalmane'. Assuming that the pharmacist knew what he was talking about, I would seem to have given a satisfactory answer to the question that had been put to me. But the answer may not seem to fit the question. The word 'how' seems to invite a response in terms of a means of doing something. 'How did you find out when the next flight departs for Rome?' admits of the response, 'By consulting the posted list of departures', or 'By asking at the information desk'. One can reply to a question of the form 'How did you find out ...?' with a sentence of the form 'I found out by ...'. But it seems that one does not respond to a question of the form 'How do you know that ...?' with a sentence of the form 'I know that by ...'.

I suppose that one could say 'I know that this capsule contains Dalmane by being told so by the pharmacist', but the fact of the matter is that one does not say such a thing, at least not in SNEAE. Instead, one would say 'I know that this capsule contains Dalmane because I was told so by the pharmacist', and perhaps one would add, 'and he ought to know'.

The same sort of seeming oddity is to be found in connection with the question 'How do you know the Chancellor?' 'I met him at a party last week' is an appropriate answer; but the question may seem to ask for an answer of the form 'By having met him at a party lask week'.

90. The fact that one must focus on here, if one is to get these matters straight, is the fact that 'know' is, indeed, an achievement verb. An interrogative employing 'how' together with an achievement verb does not require a response in terms of a means, thus one of the form 'By doing ...'.

Suppose I am to pick up George at the airport, but I do not know what he looks like. I may ask 'How am I to recognize George?' An answer might be 'By his incredible obesity', but that is not to say that

I will recognize George by doing something. 'By', in the response 'By consulting the list of flight departures', pertains to a means; but 'by', in the response 'By his incredible obesity', does not. Seeing an incredibly obese figure emerge from the plane, I walk up to him saying 'You must be George'. If he asks me how I knew that he was George, I might lie a bit and say 'By your dignified demeanour'. The truth of the matter is that I identified him as George because he fit the description I had been given. When achievement, rather than task, verbs are in question, the question 'How do you ...?' does not require an answer of the form 'By ...', where 'by' pertains to a means.

It should be noted that various nominalizations are available if one is called upon to answer the question 'How did I know that that person was George?' For one could reply, not only by saying 'I knew that he was George because he fit the description I had been given', but equally as well by saying 'On the basis of the fact that he fit the description', or by 'by his fitting the description', or by 'by his incredible obesity', and so forth.

91. What is of critical importance here is simply this: an acceptable response to the question 'How do you know that p?' may consist simply of a statment to the effect that certain conditions are, or were, satisfied.

For no matter what form my answer to the question, 'How did I know that he was George?', takes, no matter what nominalization I elect to employ, any acceptable answer will be an acceptable answer solely in virtue of the fact that it indicates that appropriate conditions were satisfied. Saying that I knew that he was George because he fit the description, or by his obesity, or by his being obese and so forth all indicate that certain conditions were satisfied.

92. If I say 'I know that p', what conditions must be satisfied for what I say to be true? The first condition that almost anyone thinks of, of course, is that p be true. And that certainly seems right.

But, perhaps, it should be noted, if only in passing, that there are

occasions on which someone says something of the form 'I know that
p' and one knows perfectly well that p is not true, and yet one would
not think that what had been said was untrue. For example, someone
says 'I just knew it, I felt it in my bones, that she would be there when
I got home, but, of course, she wasn't: am I not a fool?' Or someone
says 'Being a child was a frightening experience: going to bed at night,
I would know that there was a tiger underneath my bed, waiting to
pounce on me, but I was always too frightened to look'. Or again, a
revolutionary, facing a firing squad, says 'I know that history will
vindicate me'. History may or may not vindicate him, but whatever
the truth may prove to be, it will, no matter what, be irrelevant to what
was said. (Obviously, no sensible one is ever going to conclude:
'History did not vindicate him, so he did not know what he claimed
to know'; nor is any sensible one ever going to conclude 'History did
vindicate him, so perhaps he did know what he claimed to know'.)
Apparent counterexamples to the claim that 'I know that p' is true
only if p is true are just that, merely apparent. In all such examples,
the word 'know' is being used in special, and here irrelevant, senses.

93. If I say truly 'I know that p' then p must be true. What else is
required? The received opinion seems to be that I must believe that
p. But I certainly do not believe that.
The word 'believe' is much abused by philosophers. Although it may
be stating the obvious, one should not ignore the fact that the verbs
'think' and 'believe' are not synonyms; the contrasts between the two
provide a revealing perspective on 'believe'. One may say 'I think
about George', but one does not say 'I believe about George'. One
says 'I believe George', but not 'I think George'. One says 'I believe
in God', but not 'I think in God'. Some say, perhaps naively, 'Seeing
is believing', but I have never heard anyone say 'Seeing is thinking'.
One says 'Believe me: the Dodgers are going to win', but one doesn't
say 'Think me: the Dodgers are going to win'.
The nominalization of 'think' is 'thought'. The nominalization of
'believe' is 'belief': there is no good reason to attempt to equate beliefs

with thoughts. If I say 'The thought has just occurred to me that I had better stop work and get some exercise', I am not saying anything that could be expressed by saying 'The belief has just occurred to me that I had better stop work and get some exercise'. One may be struck by the thought that one is working too much. One is not struck by the belief that one is working too much. 'Thoughts come and go': yes, that's the way it is. 'Beliefs come and go': only if things are somewhat unstable.

94. The contrary view has been expressed: "One's repertoire of beliefs changes at least slightly in nearly every waking moment, since the merest chirp of a bird or chug of a passing motor, when recognized as such, adds a belief – however trivial and temporary – to our fluctuating store".[39]

Such a view displays a confused, and confusing, admixture of truths about beliefs and truths about thoughts, as well as untruths about either. (It is as implausible to speak of 'a repertoire of thoughts' as it is to speak of a "repertoire of beliefs". Nor can one take seriously talk about a "store" of beliefs, unless one is prepared to countenance intrinsically indefinite collections.) One might say 'I suddenly realized that the sound I heard was the sound of a passing motor', but one does not say 'I suddenly believed that the sound I heard was the sound of a passing motor'. If realization implies belief, and if the realization is sudden, isn't the belief bound to be sudden too? But if it is, why doesn't one say so?

In pondering the meaning of words it is always essential to pay attention, not only to linguistic environments in which the word occurs, or does not occur, but also to its cognates. One can express one's belief, but one can also express one's disbelief: there is no parallel cognate for 'thought'. That is something that wants explaining in any account of beliefs. Furthermore, the fact that one says 'I believe George', but one does not say 'I think George' is relevant here. 'I believe George' is an obvious transform of 'I believe what George says'. Why does 'believe', but not 'think', have such a transform? Quine and Ullian say that believing "is a disposition to respond in certain ways when the appropriate issue arises".[40] Such an hypothesis not only fails to account for the data, it creates considerable uncertainty. For example, do I, at this moment, believe that 174526 multiplied by 23 equals 4014098? If I am asked to respond to the question immediately, I should have to say that I don't know. But if I am given a few moments, I would say that, having computed it, I do believe that 174526 multiplied by 23 equals 4014098. But was that one of my "fluctuating store" before I did the computation? If one is asked whether one believes something, one may respond after a split second, or after a minute, or a month or a year by saying 'Yes, I believe that'. Was it my belief prior to considering the question, or did it become my belief, perhaps in a split second, or perhaps only after a year?

It is not, however, difficult to find an hypothesis that will account for the linguistic data, and without creating such uncertainties. A more plausible hypothesis is that, if I believe that p, I have paused, if only for a moment, to put to myself the question whether or not p is the case, and I have come up with an affirmative answer. On such an hypothesis, one sees at once why one does not say things like 'I suddenly believed that the sound I heard was the sound of a passing motor', or 'The belief struck me that ... '. For, when things are sudden, when one is struck by a thought, there is not time to pause, not time to raise a question. Furthermore, it is then not at all surprising that the verb 'believe' has the two cognate forms, 'belief' and 'disbelief'; for if I pause to put to myself the question whether or not p is the case, I may arrive at either a belief, or a disbelief, or possibly neither. And still further support for the hypothesis in question is to be found in considering the following sort of case: given Reagan's economic policies, one is, perhaps, inclined to say 'Reagan thinks rich people are more important than poor people', but one is not equally inclined to say 'Reagan believes that rich people are more important than poor people' unless one has some evidence that Reagan has actually considered the question whether rich people are more important than poor people.[41]

Despite all this, it may be objected that we frequently attribute beliefs to people without there being any evidence that they have ever considered the matter in question. For example, given that I believe that 2 is the successor of 1, and 3 of 2, and, as one is inclined to say, and so on, one may be inclined to say of me that I believe that 29,723 is the successor of 29,722: yet there need be no evidence that I have ever put to myself the question 'Is 29,723 the successor of 29,722?' But, given that I have never considered the question, such an attribution would be a mistake.

I think that one can say that if I believe that 2 is the successor of 1, and 3 of 2 and so on, that suggests that I would unhesitatingly assent to the claim that 29,723 is the successor of 29,722. But, if so, all that that establishes is that I would believe that 29,723 is the successor of 29,722 were I to consider the question. If one were to suppose that, since I would assent to the claim after considering it, I have the belief prior to considering it, one would have to conclude that I have infinitely many beliefs in my "fluctuating store"; for I believe that $n + 1$ is the successor of n, for any positive integer n. (And perhaps here one should recall the ancient rhyme: "Round and round they stared, and still the wonder grew, how one small head could contain all he knew".)

One's beliefs no doubt change in time, but not "in nearly every waking moment", unless one is mentally deranged.[42]

95. The received opinion seems to be that if I say truly 'I know that p' then I must believe that p: but that is plainly untrue.

That 'I know that p' may be true even if it is not the case that I believe that p is, first of all, indicated by admissive uses of 'know', in particular, those instances in which a person admits to knowing that p and yet explicitly professes not to believe that p. But, secondly, it is also indicated by the rather different restrictions to be found in the uses of the verbs 'know' and 'believe'. One says 'He knew at a glance that she was dead', but one does not say 'He believed at a glance that she was dead'. One might say 'When her husband burst into the bedroom, George knew at once that the game was up', but one does not, at least, not in such a situation, say 'When her husband burst into the bedroom, George believed at once that the game was up'.

96. The claim, that if one says truly 'I know that p' then one must believe that p, achieves some limited plausibility only when one restricts one's attention to claimative uses of 'know'. Neither admissive uses nor reportive uses, such as 'He knew at a glance that she was dead', suggest that if one knows that p then one must believe that p; on the contrary, they suggest that beliefs are irrelevant. But in claimative uses of 'know', other factors are brought into play.

If someone explicitly claims to know that p, it is plausible to suppose that he has paused, however briefly, to put the question to himself whether or not p is the case; in consequence, it is plausible to suppose that he believes that p. That claimative uses of 'know' implicate 'believe', while admissive uses do not, indicates, not that knowing and believing have any significant connection, but rather that claiming and believing have some significant connection, but who would have thought otherwise?

97. Suppose Josef claimed to know that p and p proved to be true. (Add, if you wish, that he believed that p.) Does it follow that he knew that p?

Any sensible epistemologist, indeed, any sensible person, would say of course not. But, unfortunately, the contrary claim is often to be heard. Seeing that a horse named 'Josef K' is running in the seventh race at Belmont, Josef bet his last dollar on Josef K to win. The horse won, at odds of 40 to 1. Afterwards, Josef jubilantly exclaimed 'I knew he would win, I knew it!' Did he know it? Not unless knowledge is simply a matter of luck. But if Josef didn't know what he claimed to know, why didn't he? After all, he was right in what he claimed. What was lacking?

A familiar answer is one couched in terms of evidence: either Josef didn't have evidence, or enough evidence, or sufficient evidence, or conclusive evidence and so forth. But though talk of evidence is sometimes relevant in evaluating a claim to knowledge, it isn't always so.

98. Evidence, as the etymology and the morphological structure of the word suggest, pertains to that which is seen, or more generally, perceived.

A scar on the surface of a cliff could be evidence that the cliff had been scaled. George's esoteric brand of cigarettes in the ashtray could be evidence that he had been there.

Here one should take care not to confuse seeing George's cigarettes in the ashtray with the cigarettes being in the ashtray: the former would be a reason to believe that George had been there, whereas the latter would be evidence that he had been there. So one says 'Seeing his cigarettes in the ashtray, I had reason to believe that George had been there'; whereas one says 'The fact that his cigarettes were in the ashtray was evidence that George had been there'. Reasons correlate with (primarily human) acts, whereas evidence correlates with facts.

99. George and Josef are walking along the beach of what they take to be a deserted island; George suddenly espys a fresh human footprint in the sand and realizes at once that there is someone else on the island. However, in his excitement, he clumsily falls on top of the print, completely obliterating it. Turning to Josef he exclaims 'There is someone else on this island, I know it!' Josef replies 'Don't be ridiculous: there's no evidence of there being anyone here'. There was evidence, but now there isn't.

Should one conclude that George did know, but now doesn't know, that there is someone else on the island? Obviously the more sensible conclusion is that if George did know that there was someone else on the island, he still knows it, despite the fact that he now has no evidence to appeal to to support his claim. In such a case, he could support his claim to knowledge, not by pointing to the evidence, for there is none to point to, but by citing a reason; thus he could say 'I know because I saw a fresh human footprint, which has since been obliterated'.

100. If a traffic officer asks me to identify myself, I might tell him that my name is Paul Ziff. Should he ask for documentary evidence, I might not be able to furnish any; possibily I lost my driver's license, car registration, in fact all my papers. Should he, for some unknown reason, say to me 'You're not Ziff, you only think you are', I might reply 'Don't be absurd: I am, I know that I am Paul Ziff'. To 'What's your evidence?', I would reply 'I have none'. Even so, I know what my name is: I don't suffer from amnesia; I am not subject to peculiar delusions and so forth. Evidence of what my name is might be useful in proving to someone else what my name is, but I know what my name is without consulting evidence of any kind.

Should I say that I know what my name is because I remember what my name is? That would certainly be an odd use of 'remember'. Suppose I am in the process of nailing a piece of wood to a door; as I am about a drive a nail in with a hammer, someone asks me 'Do you know what you are doing?' I reply 'Why certainly, I am about to drive this nail in'. If he were then to say 'What's your evidence?', I might begin to think that either he or I were mad. I could, of course, wave the hammer I have in hand at him, saying 'Just look at this!' Possibly in a court of law, the fact that I had a hammer in hand could be offered in evidence in support of the claim that I was about to drive a nail in. But I know what I am about to do without attending to the hammer I have in hand. Conceivably an absent minded professor might come out of a trance like state to find himself, hammer in hand, bent over a nail, and wonder what he was up to. And, conceivably, looking at the hammer in his hand, he might say to himself 'Evidently I am about to drive in a nail'. Or, possibly, shaking his head, he might say 'Now I remember: I was about to drive in a nail'. Fortunately, nothing of the sort has ever happened to me: when I am about to drive in a nail, I know that I am about to drive in a nail; I do such things carefully, since I have an enormous distaste for crushed thumbs. How do I know that my name is Paul Ziff? How do I know, when I am about to drive in a nail, that I am about to drive in a nail? Evidence evidently has no role to play in an answer to these questions.

101. Knowing what one's name is and knowing what one is doing are not, perhaps, exciting examples here, since one rarely is in a situation in which one would actually claim to know such things. But such situations do in fact, albeit on rare occasion, occur; they must be accounted for in any adequate account of knowledge. But claims to knowledge, that are in no way based on appeals to evidence, are not at all uncommon.

Someone says to me 'Let's go skiing in Aspen', and I reply 'No way, I can't bear snow: I dislike being in contact with anything cold'. If he were to answer (and, oddly enough, I have actually been given such an answer) 'You just think you don't like snow: snow is wonderful', I would reply (and have replied) 'I know that I don't like snow, there's no doubt about it'. It is not always true that one knows what one likes, but sometimes it is, and sometimes there is no doubt about it: I don't like snow and that is as clear as can be. How do I know that I don't like snow? More to the present point, do I know that I don't like snow on the basis of evidence? I have been in snow all too often in my life; I greet its presence with no enthusiasm whatever; I have rejoiced at seeing it melt away in the sun. I recall these experiences: do these recollections constitute evidence on which I base my claim that I know that I dislike snow? To say so would, I think, be a simple, and simpleminded, misuse of the word 'evidence'.

At a zoo, a sign on an enclosure says 'Sheep'; pointing to an animal inside staring out at us, I say to a friend 'That's not a sheep'. If he asks how I know that, I might direct his attention to certain evidence by saying 'Look at the shapes of the horns, the beard, the distinctive coloration and the shape of the head; that's an aoudad, sometimes called "the wild Barbary sheep", but in fact it is neither sheep nor goat: it is classed *ammotragus lervia*'. But if I am then asked how I know that, if what is wanted is evidence, I could offer none: I don't go about carrying text books on zoology. A sensible answer to the question would be 'I learned that years ago studying zoology'.

And to take the easiest example of all, what does evidence have to do with my knowing that $2 + 2 = 4$?

Still another example worth attending to is to be found in connection with knowledge of the position of one's body. One may know the position of one's own body without making observations of any sort, without any attention to evidence. Thus I may know that I am seated without having to look down to see the position of my legs. This, of course, need not always be so: if I were anaesthetized, or if I were in a spacecraft, orbiting in a weightless state, I might not know whether I was seated without performing some tests, or making some observations. And it is frequently the case that one does not know the position of one's hands, arms and legs if one is engaged in a complex activity such as playing tennis or performing a gymnastic stunt. Watching a video instant-replay, players often learn, to their dismay, that their knees were not bent, their wrist was slack and so forth.

102. Suppose Josef claims to know that p, p is true, Josef believes that p, and Josef has conclusive evidence that p: does it follow that Josef knows that p? That it does not can be seen by considering the following cases.

Josef believes that the butler is guilty; the butler is, in fact, guilty; furthermore, Josef has in his possession absolutely conclusive evidence that the butler is guilty. The evidence is a document, in Sanskrit, which establishes that the butler is guilty: the difficulty is that Josef cannot read Sanskrit.

To make it more plausible to suppose that Josef knows what he claims to know, let us now suppose that Josef has some knowledge, but an imperfect knowledge, of the language. And let us further suppose that Josef's belief that the butler is guilty is based on his perusal of the document. Then, even so, he need not know what he claims to know. For what if, owing to his imperfect knowledge of the language, he misunderstood what he read, but, as it so happened, his mistakes accidently cancelled out, so that he accidently arrived at the right conclusion? For example, one passage of the document, which actually stated that the chef was in the kitchen, Josef misunderstood as stating that the butler was in the kitchen; whereas another passage, which actually stated that the crime occurred in the library, Josef misunderstood as stating that the crime occurred in the kitchen: so, in essence, Josef blundered into the truth, was right, for altogether wrong reasons.

This sort of case actually occurs, in fact, frequently occurs, in connection with the use of what is, without doubt, an infallible method, in the only sense in which there are infallible methods. A student, asked to determine whether an expression having eight variables is a tautology, may use the truth table method. A truth table is an effective algorithm for determining whether or not an expression is a tautology. But a truth table with eight variables is constituted of a befuddling array of the letters 'T' and 'F'. It is all too easy to put a 'T' where an 'F' should be, and then an 'F' where a 'T' should be: in consequence, one may make a series of mistakes; but it can happen that the mistakes cancel out, that one arrives at the right answer nonetheless. Infallible methods are perfectly fine, but there is no guarantee that fallible humans (or fallible automata) make infallible use of infallible methods.

103. To salvage Josef's claim to knowledge, one has to add the proviso that, not only is Josef's belief based on his understanding of the evidence, but it is based on a correct understanding of the evidence. (Or if an infallible method was employed, then the method was correctly employed.)

We wish to say that Josef knows that p if certain conditions are satisfied, among them being the condition that he has arrived at his belief on the basis of a correct appraisal of conclusive evidence. But what is a correct appraisal? Given that the evidence is conclusive evidence, a correct appraisal must, surely, be one that leads one to the truth: nothing short of that will do. What all this indicates is this: one can say that, if Josef believes that p on the basis of a correct appraisal of evidence that establishes that p, then Josef knows that p. But to say this is not to say very much. For, first of all, as I have already argued, knowing need not be a matter of believing, and, secondly, it need not be a matter of evidence and, hence, *a fortiori*, it need not be a matter of correctly appraising evidence. Given that Josef's claim is in accordance with the indicated conditions, one can say that Josef does know what he claims to know; but such a case is merely a special case of knowledge: it cannot serve to explicate the general conception of knowledge.

104. Since what I am saying is clearly contrary to current views, it may help to attend to one such, that of Robert Nozick.[43] Nozick claims that

A person knows that p when he not only does truly believe it, but also would truly believe it and wouldn't falsely believe it. He not only actually has a true belief, he subjunctively has one. It is true that p and he believes it; if it weren't true he wouldn't believe it, and if it were true he would believe it. To know that p is to be someone who would believe it if it were true, and who wouldn't believe it if it were false.[44]

Suppose that p is true, and assume that I know that p. Then, on Nozick's account, it follows that I wouldn't believe that p if p were false. Wouldn't I? Doesn't that depend on what sort of person I am? I know that I am not doing time at Harvard. Let *ha* be an abbreviation for 'I am not doing time at Harvard'. Then I know that *ha*. But what if *ha* were not true? Can I say 'I wouldn't believe *ha* if *ha* were not true'? I very much doubt it: mere wish-fulfillment, perhaps aided by self-hypnosis, would most likely succeed in making me believe *ha*. But I know that *ha*, nonetheless. Can we patch up this view by saying that if *ha* were not true, I would not believe *ha*, given that I am not indulging in wish-fulfillment and not employing self-hypnosis? It still wont do. For what if I were drugged, or hit on the head, or tortured, or subject to a sudden aberration, or ...? One can't gloss over these three dots: the possibilities are completely undetermined: there is no available "mechanics of belief". The etiology of beliefs is an uncharted domain. It is likely to remain so for some time.

Yesterday I knew that *ha* and, at the moment, I know that *ha*; certain conditions, *c*, have been satisfied during this period. Then, can one conclude that if *ha* were not the case, but conditions *c* were satisfied, I would not believe that *ha*? This is, possibly, what Nozick has in mind. But it won't do. What if *ha* were not the case? Does it follow that I would not believe that *ha*? Only if the truth of *ha* and the satisfaction of *c* were a necessary condition for my believing that *ha*. But how could one sensibly suppose that? A fool may believe that *ha* for no reason at all. Suppose we suppose that conditions *c* include the condition that one is not a fool: but even wise men are subject to aberrations. Then let conditions *c* include the condition that one believes that *ha* only if *ha* is the case. Then it will indeed follow that if I know that *ha* under conditions *c*, if *ha* were not the case, I would not believe that it was. Such a tautology is not instructive.

One should note, if only in passing, that 'I wouldn't believe it if it weren't true' is really a silly counterfactual. I know that spiders have eight legs. But to say 'I wouldn't believe that spiders have eight legs if they didn't' would, by plain folk, and by sensible gentry, be taken to be a ridiculous claim to infallibility. Whereas to say 'I would believe that spiders have eight legs if it were true', suggests either that there is some doubt about the matter, perhaps it isn't true, or that I have achieved something approximating to omniscience.

105. Let 'giraffes have four stomachs' be abbreviated by '*g*'. If someone tells me that *g*, I may reply 'I knew that'. 'I knew that *g*' implies that I know that *g*: in such a case, 'know' has a spatio-temporal reach not to be found in connection with 'believe', or, indeed, hardly any other epistemic term.

If I say 'I believed that *g*', that does not imply that I believe that *g*; neither does 'I inferred that *g*' imply that I infer that *g*. And the same is true of 'think', 'decide', 'conclude' and so forth. Even 'I understood that *g*' does not imply that I understand that *g*, for I may be reporting an error. To find the same sort of implication as that exemplified by 'know', one must turn to terms such as 'realize' and 'aware', thus to what are, in fact, close synonyms of 'know'.

To say that 'I knew that *g*' implies that I know that *g* is not, however, to say that if I knew that *g*, it then follows that I know that *g*. Not at all: knowledge may erode in time. There are many things I once knew that I no longer know. Thus 'I knew whether or not *g*' does not imply that I know whether or not *g*. 'I knew that *g*' does imply that I know that *g*, but no such implication holds for 'know' followed by any *wh*-form such as 'whether', 'how', 'where' and so forth.

'He knew that *g*' does not imply that he knows that *g*, but it does imply that I (the speaker) know(s) that *g*: what is in play here is a speech act in which the speaker attributes the knowledge that *g*. A speaker can make such an attribution only if he has the knowledge himself. I may say 'George knew whether or not *g*' without implying that I do. I shall say that, in such a case, I am making a "nonexplicit attribution"; but if I say 'George knew that *g*', I am making an "explicit attribution". A speaker cannot make an explicit attribution of knowledge without sharing in that knowledge at the moment of attribution.

106. To glance at Nozick's account again, suppose Josef says 'George believes that *g*; George would believe it if it were true, and he wouldn't believe it if it weren't true'. Given that *g* is true, does it follow, on Nozick's account, that Josef knows that *g*?

Clearly not: for Josef could, without difficulty, continue 'Unlike George, I would believe that *g*, even if it weren't true'. Hence, assuming that Josef is right about himself, on Nozick's account, Josef doesn't know that *g*. Hence in saying what he said about George, Josef could not have made an explicit attribution of knowledge. More simply, he did not say or indicate that George knows that *g*.

107. The contrast to be found between nonexplicit and explicit

attributions of knowledge is a reflection of the diachronic and synchronic aspects of knowledge.

The knowledge that p is knowledge now. This is obvious when p has an atemporal cast. Thus if I say 'George knew that giraffes have four stomachs', I at once imply, if I am speaking literally, that I know that giraffes have four stomachs. But the knowledge that p is knowledge now even when p has a temporal cast. If I say 'George knew that his horse would lose', then I imply that I know that George's horse lost. The knowledge whether or not p can have a different temporal aspect. Knowledge that is lost, or yet to be gained, is not the knowledge that p, but the knowledge whether or not p, or knowledge why, or how, or where and so forth. One says 'I used to know whether or not p', but not 'I used to know that p' (save, perhaps, by way of laconic comment on earlier errors). Hence, it should now be clear, or anyway clearer, why expressions of the form 'know whether', 'know how' and so forth are not genuinely analysable in terms of 'know that'.

CHAPTER X

A POSITION TO KNOW

108. ALTHOUGH claimative uses of 'know' are common, disclaimative uses are equally common, and, from an epistemological point of view, equally interesting.

If someone were to ask me why the Dean of the College does the peculiar things he does, I should have to answer that I really don't know. If I were then asked why I don't know, the answer would be easy: I don't know what he's up to because, not being a member of the administration, and not being privy to his councils, I am plainly not in a position to know anything about such matters. If I am not in a position to know whether or not p then I don't know whether or not p.

Even though it follows that if I am not in a position to know whether or not p then I do not know, it does not follow that if I am in a position to know whether or not p, I then know. For example, having attended the last departmental meeting, I may well be in a position to know whether or not a certain resolution was passed; nonetheless, it is likely to be the case that I do not know, for I pay very little attention to resolutions and their passages.

109. When is one in a position to know whether or not p? To speak of "being in a position to know" is, perhaps, to speak somewhat metaphorically. I am inclined to suppose, however, that the metaphor is something of a dead metaphor, for the expression, 'in a position to', is an established idiom of SNEAE, and of English.

Whether one is in a position to know whether or not p must, of course, depend on p; for if I am in a position to know whether or not q, it does not follow that I am in a position to know whether or not p, if p and q are reasonably distinct and different. Certain conditions must

be satisfied for me to be in a position to know whether or not p; quite
other conditions may have to be satisfied for me to be in a position
to know whether or not q. But, in any case, if I am to be in a position
to know whether or not something is the case, certain generic and
fundamental conditions must be satisfied. What conditions?

110. To ask the hard question is easy, but, in this instance at least,
the form of the answer is reasonably clear: the relevant conditions can
only be those that pertain to the possibility of error.
If one knows that p then p must be the case. The truth of p is clearly
the pivotal issue in claimative uses of 'know': if I sincerely claim to
know that p, but p isn't so, then I am in error; such an error suggests
that I was not in a position to know whether or not p, and, hence,
the requisite conditions for being in such a position could not have
been satisfied: but that cannot be right.

111. If I am to know whether or not p, I must be in a position to know
that. But even if I am in such a position, it does not follow that I know
whether or not p: why not?
Let's reconsider the previous case cited. If I attended the departmental
meeting, I was in a position to know whether or not a certain
resolution was passed. I did attend the meeting; even so, I do not
know whether the resolution was passed: why not? Though I was
present, I paid no attention to what transpired.
But, perhaps, here one might argue that, since I do not know whether
the resolution was passed, I had not, in fact, been in a position to
know that. To have been in such a position, it was necessary, not only
to have attended the meeting, but to have been attentive when
present; given that I had been inattentive, I had not been in a position
to know whether or not the resolution was passed.
To accept such an argument would virtually commit one to misusing
the expression 'in a position to'. One speaks, not only of being "in
a position to know", but of being in a position to "pay a debt", "visit
the Louvre", "investigate a problem" and so forth. If George is in a

position to pay a debt, but elects instead to squander the money on a trip to Vegas, one could hardly argue that he couldn't have been in a position to pay the debt, since he didn't pay it. It is one thing to be in a position to do something, it is another to avail oneself of the opportunities of that position.

If I had attended the meeting but, owing to my inattentiveness, I had not noted whether the resolution passed, I should say that, even so, I had been in a position to know. Whereas, if I had attended the meeting, been attentive and, even so, had been unable to determine whether the resolution passed, owing to the fact that people had constantly been screaming in my ear, I should say that, even though I had been present, I had not been in a position to know whether the resolution passed. The difference is that, in the former case, I did not avail myself of the opportunities of the position, whereas, in the latter case, the opportunities were lacking.

112. There are two distinct factors in play here. If I am not in a position to know whether or not p, then I do not know. But even if I am in such a position, I may not know.

If I claim to know that p and p is true, I may still be in error, and for two quite different reasons: That I am not in a position to know whether or not p is one, that I have not availed myself of the opportunities of the position I am in is another.

What this indicates is that if I am to know whether or not p, both of these possibilities must be precluded.

113. The strongest possible demand that one could make here would be this: one could insist that one knows whether or not p if and only if one is in a position such that, in that position, there is no possibility of error with respect to the truth of p. The immediate consequence of such a requirement would be that no one ever knows anything. There is no position which a human being can be in and in which all possibility of error is precluded. The moment one opens one's mouth, the possibility of error flickers snakelike into life. 'I know that that

is my hamburg' the stockbroker says, indignantly, at someone grabbing his Homburg off the hat rack. To speak or to misspeak, to hear or to mishear, are ever available alternatives that guarantee the possibility of error.[45] Visiting an eye, ear and nose specialist, one says 'Doctor, I've a terrible pain in my left rear'. Noting his glazed look, one may correct oneself, 'I mean in my left ear'. Or again, lecturing to a class in epistemology, I say, while writing on the blackboard, 'None of you are really going to deny this', whereupon the class bursts into laughter. Puzzled, I happen to glance at the blackboard and discover that I have written 'I know that $2 + 2 = 22$'.

114. One doesn't avoid the possibility of error by indulging in fantasies about "incorrigibile statements". To make a statement, one has to open one's mouth and say something, and one can state or misstate.

I am seated in a dentist's chair: he has just given me a local anaesthetic. Prodding me with one of his instruments of torture, he says 'This doesn't hurt, does it?' and I at once reply 'Yes it does', and then, after a moment, retract what I've said, saying 'No, it didn't; I suppose I'm a bit nervous'. Did it hurt? No, but, owing to my nervousness, I misstated the matter. And again, the unfortunate soldier who has had his left leg amputated, but does not know that it has been amputated, says 'I have a horrible pain in my left ankle': he is in error, for one can't have a pain in one's ankle unless one has an ankle to have the pain in, which is not to deny that he may be in considerable pain. A phantom limb pain is not a pain in one's missing limb, but a perfectly real pain which one mislocates, for altogether explicable reasons.

DesCartes, I have been told, claimed to know that he existed, and he supported his claim by saying something to the effect that since he thinks, he must exist. (Not being an historian, I would not know whether or not that is a reasonable rendition of "*Cogito ergo sum*".) But, as DeCheval later said (but of course was ignored), one can just as well doubt whether one thinks as whether one exists. And certainly one can doubt whether one doubts: I very much doubt that DesCartes had any doubts about whether or not he existed. There's nothing sacrosanct about the conception of existing, or of being an existent being, or being a being and so forth. Nor is there anything particularly privileged about the conception, or one's conception, of thinking or of doubting. Nothing precludes the possibility of one's having a misconception of any of these matters. And if one does have a misconception, quite possibly all inferences are at once aborted, and any claim to knowledge miscarries.

115. If it were true that to know whether or not something was so, one would have to be in a position in which all possibility of error was precluded, then one would have to conclude that no one ever knows anything. But that would be tantamount to concluding that the conception of knowledge, or, more precisely and more sensibly, such a conception of knowledge, has remarkably little utility.

There is an unmistakable temptation here to say that such a conception of knowledge has no utility whatever, but that, I think, would be an overstatement. For dour types, such as myself, there is something intriguing in a conception of knowledge that allows one to conclude that no one ever knows anything. After all, we do live in dark ages; the heraldic emblem of the day can only be: Ignorance rampant in a field of folly. One could take a misologistic delight in the realization that all claims to knowledge are untrue, that the truth is that knowledge is an impossible dream.

A serious epistemic analysis of 'know' and 'knowledge', if it is to be taken seriously, must, however, manage to accord with the available data. The marginal misologistic utility of a conception of knowledge that allows one to conclude that no one ever knows anything can hardly provide an adequate explanation for the existence of, and continued use of, the epistemic terms 'know' and 'knowledge', and the corresponding terms in various natural languages. Words are sometimes likened to tools; but such a metaphor is totally inept; if one is to indulge in figure, then words are, surely, conceptual instruments, instruments that enable one to cope with the demands of a highly complex ecosystem. The fact of the matter is that claims to knowledge are often made and are often accepted by reasonable persons: the problem is to understand what is being claimed and what is being accepted.

116. Since one cannot reasonably require that, if one is to know whether or not p, one must be in a position in which all possibility of error be precluded, what can one reasonably require?
Obviously this: one must be in a position in which any possibility of error with respect to p may safely be discounted. I shall speak of such a position as a "safe position with respect to p", or more simply as a "safe position".
Although it may be altogether evident, it must be stressed that if one is in a safe position with respect to p, it does not follow that one is in a safe position with respect to q, if p and q are reasonably distinct and different. Furthermore, the difference between p and q may be ever so slight; how slight is perhaps indicated by the fact that if I am in a safe position with respect to p at one time, it does not follow that I am in a safe position with respect to p at some other time. All of us, and all things around us, are in the grip of entropy, and knowledge, as well as everything else in this world, decays in time. That I know that something is so today does not mean that I knew it yesterday, or that I will know it tomorrow.

117. The entropic degradation of knowledge that occurs in time should not be confused with certain evolutionary aspects of knowledge, not that such a confusion is likely. But one must take care not to confuse being in a position to claim to know whether or not p, with being in a position to know whether or not p.
One may be in a position to claim to know whether or not p, and yet not, in fact, be in a safe position. Of course, if one knows that one is not in a safe position, then one cannot sincerely claim to know (unless, perhaps, one is seriously confused). One is in a position to

claim to know that p only if one is in a position which is presently deemed to be a safe position with respect to p. But, owing to the evolution of knowledge, a position deemed to be safe at one time, need not be deemed to be safe at another time. For example, if in 1882 one knew that George was 25 years old and manifested the attributes of a typical 25 year old, then one was in a position to claim to know that George's identical twin, Giorgio, was also 25 years old and manifested the attributes of a typical 25 year old, unless he, unlike George, had been stricken with some disease, early senescence and so forth. But the safety of the position to claim to know that has, owing to the emergence of relativity theory and the possibility of space travel, eroded in time. Tomorrow, perhaps, no one will be able to make any inferences from knowledge about the age or appearance of one twin to the age or appearance of the other, without knowing whether they have been engaging in space travel at velocities close to the speed of light.

Here one can see precisely what is wrong with Ayer's definition of knowledge in terms of "the right to be sure". One has the right to be sure that p, when one claims to know that p, if one is in a position that is at present deemed to be safe with respect to p. But since what is, at present, deemed to be a safe position need not, in fact, be a safe position, one need not know what one claims to know, even if one has, at present, every right to make such a claim.[46]

118. The possibility of the evolution of knowledge, though no doubt undeniable, is something that wants explaining. For how is it possible to discover that one is mistaken in a claim to knowledge?

Here one must turn to cases. George tells Josef 'My car is in the garage'. Josef replies 'I don't think so'. Irritably, George reponds 'Don't be stupid! I know that the car is in the garage', whereupon Josef leaves to look. Returning, Josef says 'You're wrong; I told you so: it isn't there'. Suppose George then proceeds to the garage, looks about and fails to find his car. He might then turn to Josef saying 'You

were right and I was wrong: the car isn't here'. Thus George has
retracted his claim to know that p: but why?

What has in fact transpired in this seemingly simple situation is, in
fact, rather complicated.

119. I shall say that, when George first told Josef that the car was in
the garage, George was in position $Sg1$, whereas Josef was in position
$Sj1$. When Josef returned, after taking a look, he was then in position
$Sj2$, whereas, after George went to the garage himself, George was
in position $Sg2$. When Josef was in position $Sj1$, he did not deem
himself to be in a position to know whether or not p, but once he was
in position $Sj2$ he did. On the other hand, when George was in
position $Sg1$, he mistakenly thought that he was in a position to know
whether or not p, and then, when he was in position $Sg2$, he again
thought that he was in a position to know whether or not p, only then
he arrived at the opposite conclusion with respect to the truth of p.
What all this implies is this: Josef judged that, with respect to the truth
of p, $Sj2$ was a safer position than $Sj1$; whereas George, when in $Sg1$,
judged $Sg1$ to be safer than $Sj1$, and then, later, judged that $Sg2$ was
a safer position than $Sg1$, and, possibly, from the perspective of $Sg2$,
judged that $Sj1$ was a safer position than $Sg1$.

One can discover today that one was wrong yesterday only if, today,
one is in a safer position than one was in yesterday. Judgements of
the relative safety of positions are absolutely fundamental and
essential in arriving at sound epistemic conclusions. But, unfortunately, nothing guarantees the correctness of such judgements.

120. From this vantage point, one can see that the skeptics who
refused even to look through Galileo's telescope have possibly been
mistakenly maligned.

Presumably, they deemed themselves to be in a safe position with
respect to the truth of p, say position Si. To have looked through the
telescope would have been to put themselves into another position,
Sj; but, being in position Si, they thought they had good reason to
suppose that position Sj was less safe than Si; hence, what they could

see, were they to look through the telescope, would either confirm what they already knew, or, if it did not, could be explained away as images supplied by the Devil.

121. If someone sincerely claims to know that *p*, but *p* is not the case, then, assuming that he was in position *Si*, one can conclude that position *Si* is not a safe position. Given that the claim was a sincere claim to knowledge, a mistake in judgement may have occurred. We report such a mistake by saying of the person, 'He thought he knew that *p*'. Furthermore, if *Si* is not a safe position, then, even if p is the case, one who claimed to know that *p*, when in *Si*, did not know what he claimed to know. And, again, one might say of him 'He thought he knew that *p*'. But, if one is not careful, one can easily be confused by such a report.

Cook-Wilson claimed that there was a patent absurdity here; for someone who sincerely claims to know is not claiming to think he knows: a man who claims to know "does not say to himself 'I think I know', for that must mean he knows he does not know".[47]

I am inclined to think that the spirit of Cook-Wilson's claim is quite right, even though the letter of it is not. If what is in question is, say, the answer to a mathematical problem, one certainly can say 'I think I know, but I'm not sure: give me a minute or two to think about the matter'. But if I sincerely and explicitly claim to know that *p*, I am not claiming to think I know that *p*: I am claiming to know that *p*. Why, then, if I discover that I was in error, do I report this, either by saying 'I thought I knew that *p*', or by saying 'I thought that *p*'? (Which I am apt to say depends on the error: if *p* is true, but the position unsafe, I am apt to say the former; if *p* is not true, the latter. Of course, if what I discover is that I have been subject to, and the error is attributable to, a conceptual confusion, I might say something like 'I thought it made sense to say 'I know that *p*', but now I see it doesn't'.) Or why is one, who knows that I am in error, apt to say to me 'You only think you know that *p*, but you don't'?

One can readily resolve Cook-Wilson's dilemma by attending to the difference between what I think, or what I judge, when I claim to know that *p*, and what I claim when I claim to know that *p*. If, in position *Si*, I claim to know that *p*, then I think, or judge, that *Si* is a safe position, but, unless I am claiming to know that I know that *p*, I am not claiming to know that *Si* is a safe position; I think, or judge, it to be so, otherwise I could not sincerely claim to know that *p*. If, having claimed to know that *p*, and having discovered that I was mistaken, I then admit this by saying 'I thought I knew that *p*', what I am admitting is, not that there was a time at which I was prepared to say 'I think I know that *p*', but that there was a time at which I thought that *Si* was a safe position.

122. Cook-Wilson's claim, that a man who claims to know "does not say to himself 'I think I know', for that must mean he knows he does not know",[48] deserves further consideration here, for it is symptomatic of a peculiar difficulty that epistemologists seem to suffer from. But, of course, not only epistemologists: philsophers in general, and linguists in particular, seem remarkably susceptible to this ailment. In contrast with Cook-Wilson, I see no difficulty whatever in supposing that a man who claims to know may say to himself 'I think I know'. And it may certainly seem that, in saying this, I am disagreeing with, and even contradicting, Cook-Wilson's claim. But I do not think so.

If someone asks me if I know where Bentham discussed the problem of defining the word 'right', I might reply 'I think I know, just wait a moment'. And then, after a moment's thought, I might say 'Yes, I do know: see his *A Fragment on Government*'. If Cook-Wilson were up and about today, instead of in the grave, would he object to what I have just said? I don't know. But the point of view would, I am reasonably sure, seem very odd to him. For, unlike Cook-Wilson, I view language and discourse from a diachronic evolutionary point of view. Does one say 'I think I know' in making a claim to know? Yes, in the course of making such a claim, when making such a claim is viewed as a process requiring time, subject to growth, evolution and decay. But if one adopts a synchronic stance, 'I think I know' can seem absurd. If, in reply to 'Do you know whether George is at home?', I reply 'Yes, I think I know that George is at home', I am likely to provoke the response, 'Well, do you know or don't you?'

123. Synchronicism is (or has been) the chronic ailment of philosophic linguists. Or, at any rate, it provides the kindest explanation one can give for claims to the effect that children have an "innate knowledge" of grammar.[49] One can attribute innate knowledge to living organisms only if one ignores the causal and temporal differences between conception and maturation.

One can sensibly claim that children have an innate capacity to see; but one cannot sensibly claim that children have an innate view of their environment. Yet the claim that children have an innate knowledge of grammar is stricly on a par with the latter absurdity. (And so one could imagine a "universal view", just as some imagine a "universal grammar".)

Even though children have an innate capacity to see, infants kept in complete darkness for an extended period will prove to be blind: owing to a lack of stimulation, the optic nerve will atrophy. What is in question is the maturation of neural mechanisms, and such maturation is a function, not only of innate neurophysiological structures, but of the environing ecosystem as well.

It is conceivable that, as is claimed, children are innately determined to construct grammars of a certain sort; but even if that is true, whether or not a child will ever achieve a knowledge of grammar depends, not only on innate structures, but on the environing ecosystem. Feral children may never learn to speak a language, may never achieve a knowledge of grammar: being nurtured by wolves is not, or so the available evidence suggests, conducive to the maturation of the neural mechanisms requisite for achieving a knowledge of grammar or language.

If children actually did have an innate knowledge of grammar, then one would have to conclude, either that feral children have a knowledge of grammar, even though they fail to display any such knowledge in any way whatever, or that feral children must all be genetically defective. One might as well conclude that we're all innately dead: for any organism, after a brief period of being subjected to random stimulation, ends up dead.[50]

124. Synchronicism is also implicated in the frequently made mistaken claim that if I know that p, then I know that I know that.[51]
To know that I know that p, assuming again that I am in position Si, then not only must Si be a safe position with respect to p, but I must know that Si is such a safe position. Let 'q' be 'Si is a safe position with respect to p'. Even if Si is a safe position with respect to p, there is no good reason, indeed, no reason, to suppose that Si must be a safe position with respect to q.

125. Though they are not too common, there are occasions on which one claims to know that one knows. For example, if I am asked a difficult question, I may reply 'I know the answer, but you'll have to give me time to mull it over'. After a while, not having answered the question, I may be told 'You don't know it', to which I may respond by saying 'I know that I know the answer; just wait a bit'.
It must be noted, however, that it is not as easy to describe plausible cases in which one claims to know that one knows that p. If one says something of the form, 'I know that I know ...', plausible completions of the utterance are provided by 'the answer', 'where he has gone to', 'how to solve it', 'what he said', 'which turn to take'. Suppose, for example, I do indeed know which turn to take, and suppose that the turn to take is the left turn. Why, then, is it plausible to say 'I know that I know which turn to take; just wait and see'; but one is not, in fact, apt to say 'I know that I know that the left turn is the one to take'? Furthermore, should I prove to be in error, I might say 'I thought I knew that I knew which turn to take', but one does not say, and one is not apt to say, 'I thought I knew that I knew that the left turn was the turn to take'.

The obvious answer is that what is at issue when one says 'I know that I know which turn to take; just wait and see', is not what would be at issue if one were to say 'I know that I know that the left turn is the one to take'. For, in the first case, one is primarily concerned to determine whether I know which turn to take, and such a question may be of considerable practical importance. But, in the second case, what would be at issue would be of interest only to someone concerned with epistemological matters. But there is more to it than that.

If I claim to know that I know which turn to take, I am claiming to know that I am in a safe position with respect to knowing the solution to a problem. But the knowledge involved in knowing the solution to the problem may be akin to the knowledge involved in knowing how to do something: it may be internalized. Such knowledge can be displayed in time, manifested by behavior: 'I know that I know which turn to take; just wait and see'. If one were to ask me how I know which turn to take, I might reply that I don't know how I know that, but, nonetheless, I know that I know which turn to take. Whereas, if I were claim to know that I know that the left turn is the turn to take, I would be claiming to know that I am in a safe position with respect to knowing that the left turn is the turn to take: such knowledge is, and must be, explicit: here one is virtually compelled to adopt a synchronic stance.

126. Although what is in question is knowing that one knows that p, the following conclusions, that one is driven to when one considers the following kind of case, are significant.

Suppose Josef knows George, and George is the Chancellor of the University, but Josef doesn't know that George is the Chancellor. Sidney asks Josef to speak to the Chancellor on his, Sidney's, behalf. Josef replies that he would be happy to oblige, but he doesn't know the Chancellor, and wouldn't it be better to ask someone else to intercede? Whereupon Sidney replies, angrily, 'Josef, if you don't want to help me, just say so, but don't lie to me: I saw the two of you having dinner together last night'. Josef answers, indignantly, 'I had dinner with George last night'. 'George is the Chancellor', shouts Sidney, in a rage.

Does Josef know the Chancellor? Evidently. Did he, prior to this conversation with Sidney, know that he knew the Chancellor? Evidently not.

127. It has often been pointed out that if someone believes that p, and q is equivalent to p, it does not follow that that person believes that q. Thus, again, if George is the Chancellor, but Josef doesn't know this, Josef may believe that George is a superior being, and yet not believe that the Chancellor is a superior being. Given that Josef believes that George is a superior being, can one conclude that Josef believes that he believes that George is a superior being? It seems altogether possible that Josef both believes that George is a superior being and yet does not believe that he has any such belief. People can deceive themselves, lie to themselves. Perhaps Josef is a professed egalitarian; he despises the thought of anyone's being a superior being. But, though he will not admit it, even to himself, he really believes that George is a superior being. He doesn't believe that he has any such belief, but he may have, nonetheless.

But if this is granted, namely, that one can believe that p, and yet not believe that one believes that p, some warrant is required to accept the conclusion that, if one knows that p, then one knows that one knows that p. For although there are obvious differences between 'know' and 'believe', what difference(s) would serve to account for this difference? The answer is: none. To say otherwise is to make mysteries where there are none.

128. Since very clever people have maintained that if one knows that p, then one knows that one knows that p, it may help to say a few more words about this.

I would have thought that, for any academic epistemologist, one who teaches at a university, the truth, that one may know that p, and yet not know that one knows that *p*, would be altogether obvious from classroom experiences. I am quite sure that my students know all sorts of things that they vehemently deny knowing, being, as students are wont to be, of a skeptical frame of mind. I am not the least bit persuaded that they do not know the things they profess not to know; but I have no doubt whatever that they do not know that they know these things. A student who, in the classroom, may insist that he not only does not, but cannot, know whether he has any money on him, will, nonetheless, outside of the classroom, assure me that he has enough money to buy us both drinks. If I question his claim by saying 'Are you sure you have enough?', he may cheerfully reply 'Don't worry, I know I've at least ten dollars on me'. Does he know that? Perhaps, but he certainly doesn't know that he knows it. Students who, though accomplished in mathematics, deny that they know that 2 + 2 = 4, do not, in so doing, betray any ignorance of arithmetic: they merely display their ignorance of a sound epistemology.

But there is a fragment of a truth in the claim that if one knows that *p*, then one knows that one knows that *p*, which becomes apparent when one reflects on the attitudes displayed by skeptically minded students. For if one is prepared to say of oneself that one knows that *p*, then one is not, then and there, in a position to deny that one knows that one knows that *p*. Thus, if I say 'I know that *p*', and someone then queries me with 'Do you know that?', what am I to say? Surely not: 'I know that *p*, but I don't know that I know that *p*'. But to conclude, on such a basis, that if one knows that *p*, then one knows that one knows that *p*, would be to confuse the implications of speech acts with the facts. 'I knew that *p*' implies that I know that *p*, but that doesn't mean that if I knew that *p*, I now know that *p*. Just so, 'I know that *p*' implies that I am, or may be, prepared to state 'I know that I know that *p*', but that doesn't mean that if I know that *p*, then I know that I know that *p*.

129. I am now in a position to tender, and we should now be in a position to consider, the following hypothesis: One knows that *p* if

and only if p is true, and one is in a position such that, in that position, any possibility of one's being in error with respect to the truth of p may safely be discounted.

That such an hypothesis poses considerable problems is, or should be, fairly obvious. But there is one immediate difficulty that may not meet the casual eye. On the hypothesis in question, one knows that p if certain conditions are satisfed; the conditions pertain to truth and the possibility of error. But what does truth and the possibility of error have to do with knowing? I know that the question seems odd, but it is something that one must think about.

CHAPTER XI

ANALYSIS

130. ONE SPEAKS of knowing a variety of matters: one may know George, that giraffes have four stomachs, how to swim, the proof, a way to do it, the answer, what the problem is and so forth. Knowing that p, on virtually all accounts, requires that p be true. But neither knowing George nor knowing how to swim seem to have anything to do with truth. Why not?There is an easy answer to this question that one should not, cannot, accept unless one is driven to it as a last resort, namely, that 'know' has a multiplicity of meanings and this remarkable polysemy is activated by the diverse linguistic environments in which the verb occurs. The principle being appealed to here is what I have elsewhere called "Occam's eraser": do not multiply dictionary entries beyond necessity.[52]

A more plausible view, and, I believe, the correct one, is that the various uses of 'know' exemplify, not different meanings, but different senses of the word. The meaning of a word can be characterized as a set of conditions associated with the word. When one uses the word (assuming that one is not using the word in a "nonce" sense), some subset (or, if tropes or discourse operators are in play, some subset of a superset,) of this set of conditions is invoked. The particular subset of conditions invoked in a given use of the word constitutes the sense of the word in the given case.

An adequate analysis of a term requires one to account for all senses of the term that are not attributable to standard tropes, discourse operators and the like. Thus semantic analysis aims at a characterization of the meaning of a term. Sense is, so to speak, a matter of direction: one better knows where a term is going if one knows where it came from.

121

131. If I know that p, then p must be true. But if I know George, not truth, but existence becomes relevant. In what I have called the "acquaintance" sense of 'know', one cannot know George if George is dead: 'know' then converts to 'knew'. How are we to understand this switch from truth to existence?

To proceed by indirection, consider the differences between knowing how to open a certain window and knowing how to swim. If I know how to swim then I have learned to swim. But if I am asked 'How did you know how to open that window?', I might reply 'I just figured it out a moment ago'. Figuring out something is more akin to solving than to learning. 'How did you know that the butler did it?' – 'I figured it out from the clues: he was in the library, but the serving tray was still in the kitchen'. Figuring out and solving may lean in the direction of the truth, but learning need not.

But all this seems to make matters worse: one moves from what one knows to the truth of p to the existence of George to learning to swim by a hop, a skip and a jump. That there be such a jump, however, should not be altogether surprising: it is suggested by comparative linguistic evidence. Though one translates 'I know how to swim' into Italian by 'So nuotare', in German one uses neither 'kennen' nor 'wissen' but 'konnen': 'Ich kann swimmen'. A similar jump, however, is to be found in the use of 'learn': one can learn to swim and one can learn that George is at home. And that this, too, is something of a jump is again suggested by comparative linguistic evidence: one translates 'I learned to swim' into Italian by 'Ho imparato a nuotare', but 'I learned that George is at home' by 'Ho scoperto che Giorgio e a casa'; one does not use the verb 'imparare' in such a case. Such comparative linguistic data, however, is merely suggestive: English is not German and not Italian.

132. If one is to characterize the meaning of 'know' in terms broad enough to encompass all of its various uses, the characterization will have to be relatively abstract.

Let E be a dummy for any well formed expression that may occur in the environment 'I know ...'. Then consider the hypothesis: if I know E, then my knowing E constitutes an increase in coherence, in contrast with my not knowing E.

My knowing E can thus be said to maximize coherence. To seek to know E is, then, to seek an increase in coherence; to know E is to achieve it. I am not simply saying that 'know' is an achievement verb; I am saying that knowing E is an achievement, perhaps a modest or even trivial one on occasion and depending on what E is, but an achievement nonetheless.

133. If E is that p, then what is in question is truth or falsity. If my knowing that p is to constitute an increase in coherence, then p must be true.

Making false statements decreases coherence, for false statements fail to conform to reality. Or, to avoid even the appearance of metaphysics, a statement is not true if its truth conditions are not satisfied. When a statement is true, the truth conditions associated with the statement are identical with a set of conditions that are in fact satisfied. Making a true statement instantiates this identity and thus, in contrast with making a false statement, constitutes an increase in coherence.

If I know E, my knowing E must constitute an increase in coherence. If so, then, if E is itself irrevocably and inalterably incoherent, I cannot know E. Thus, for example, one cannot know that p and not p, and one cannot know that $2 + 2 = 5$.

There are, of course, instances in which making a false statement can serve to increase coherence. If one lies in stating that p, to complete the lie, one may have to state falsely that q and then that r. The false statement that r increases local coherence in the system of statements constituted by p, q and r. But a genuine increase in coherence is to be evaluated in the way one evaluates a decrease in entropy. A local decrease in entropy is accomplished only at the price of an overall increase: the sandy beach grows unbearably hot only at the expense of solar energy. Though the false statement that r may serve to increase the coherence of p and q, no genuine increase in coherence is achieved in making the false statements p, q and r when one views the matter with sufficient perspective.

(But what if one believed enough false statements such that coming to know certain truths might serve only to decrease coherence? (This is akin to, though not the same as, asking 'If ignorance is bliss, isn't it folly to be wise?') Here one must focus on the fact that what any individual knows stands in some relation to what is known. For one can, and one often does, quite sensibly employ an agentless passive in saying something to the effect that it is known that p. The domain of knowledge is the domain of what is known, not simply what some individual knows: one abstracts from reference to persons when one is concerned simply with what is known. Hence it would not be enough for a single individual to be coherently deluded: the coherent delusion would have to be indefinitely spatio-temporally wide-spread.[53] And what if that occured? What indeed! A given conception may or may not be of utility.[54] There is no guarantee that our current conception of knowledge is fool-proof and fail-safe.)

The case of one who knows that p but refuses to believe that p is of particular interest here. Consider a person of whom it is said 'He knows that p, but he simply can't face the fact: he can't believe it, but the knowledge is driving him mad'. Such a person is in a relatively incoherent psychological state. The incoherence, in such a case, is directly attributable to the increase in coherence constituted by his knowing that p. The case is strictly analogous to one in which, having told a series of lies, one states a truth that conflicts with the lies. The lies may have served to achieve some local coherence, but that local coherence is then decreased, or even destroyed, by the addition of a truth.

134. If I know that p, then my knowing that p must constitute an increase in coherence. But one can know that p without stating that p: where, then, is the increase in coherence?

Perhaps then I believe that p, and my belief is in accordance with reality. But what if I have never considered whether or not *p*? I may know that *p* nonetheless. Then perhaps my behavior would exemplify an increase in coherence. But what if I know the answer to a problem, but don't think I do? Then, perhaps, with appropriate prodding, I would come up with the answer. On occasion, one prods students who claim not to know the answer to a problem by saying 'You know it: just think about it for a few minutes', and often they do know it and produce it.

But the fact of the matter is that I may know that *p* without manifesting that knowledge in any way whatever. For example, a moment ago it would have been true to say of me 'He knows that the Moon is not identical with the Earth', but a moment ago it would not have been true to say of me that I had ever expressed such a thought, either directly or indirectly, or that that thought had ever crossed my mind. How, then, did my knowing that *p*, in that case, constitute an increase in coherence?

135. Here one must turn to counterfactuals. Consider a delicate piece of crystal, a wine glass. Such a glass would shatter if it were dropped an appreciable distance onto a concrete floor. That it would shatter is attributable to its molecular structure in virtue of which such an object is rightly classed fragile. Sensible counterfactuals are always based on the characteristics of what is in question.

Someone holding a match in his hand says 'This match would ignite if it were struck'; it is then pointed out that the match is wet. One can then conclude 'The match wouldn't ignite if it were struck' or 'The match would ignite if it weren't wet and were struck'. The latter conclusion may seem to smack of triviality, but it is, in fact, neither trivial nor incorrect. For one could not say the same of an ordinary wet toothpick: the toothpick would not ignite even if it weren't wet and were struck. (To achieve relative triviality, one could say 'The toothpick would ignite if it were appropriately coated with chemicals, were not wet and were struck'. That even this is not a genuinely trivial statement can be seen by reflecting on the impact such information could have on a culture in which matches were unknown.)

If I know that p, then I would state that p if I were concerned to state the truth with respect to whether or not p, and, of course, in the absence of contravening considerations (I do not speak the language, my tongue has been torn out and so forth). The case of one who knows that p, but refuses to believe that p, is analogous to the case of the wet match. Such a person would state that p if he were concerned to state the truth with respect to whether or not p, and if he were not psychologically disturbed by that truth.

136. To skip to the case in which E is George, what is then in question is existence or nonexistence. If so, existence is wanted: the non-existent is less coherent that the existent.

If something exists, it constitutes an entity of some sort, it has some sort of unity. Once George dies, and so ceases to exist, I can no longer truly say 'I know George': existence is here the analogue of truth.

Why can't one know the way to build a perpetual motion machine? The obvious answer is that there is no such thing: the way does not exist. Just so, today one cannot know where Dante Alighieri (as opposed to his dust) is, for he isn't anywhere: he no longer exists, and so there is no place where he is: it doesn't exist. And again, one can't know the reason why Reagan wasn't elected president, because he was, and hence there can't be any reason why he wasn't.

But again, there are various ways in which the nonexistent may seem to be more coherent than the existent. For one can imagine a world in which things were better arranged than they are in this one, a world in which virtue was rewarded, where misery was eliminated, a coherent utopia in short. Compare the nonexistent planet of Utopia with Earth: isn't Utopia more coherent than Earth?

It would be if it existed, but since it doesn't, it isn't. Because if one is to take the coherence of Utopia seriously, then one must work out the details. One must complete the picture: completeness is a fundamental feature of coherence.

In imagination the restrictions of coherence are relaxed, or shifted or even ignored. Hamlet is a fictitious, an imaginary, character. Did he have warts? A real person either does or doesn't (ignoring borderline problems). With Hamlet, the question does not arise, for nothing in Shakespeare's *Hamlet* bears on the question. But if one compares the nonexistent Hamlet with any existent being, even a poor psychotic displaying multiple psychotic personalities, the psychotic presents a more coherent aspect than Hamlet. For the psychotic, being real, has a history, a neurophysiological structure (but possibly a weird one), a body either with or without warts. He has a specific weight, which can conceivably be determined. What did Hamlet weigh? What did he eat, if anything, for breakfast? The limits of the answers are determined by the limits of the play, by the play of Shakespeare's imagination.

Again, within the limits of the play, one finds pockets of coherence. But from a broader perspective, the merely imaginary collapses into relative incoherence.[55]

137. I think that the view that I am urging one to adopt here will seem to many as somewhat, or even remarkably, strange. So, perhaps, it may help to elaborate a bit further.

Consider these two cases: In the first, there are two cats in my garden; one of them is my cat Osiris; the other I do not know. In the second case, there are again two cats in my garden; one of them is my cat Osiris, the other is my cat Tao. Thus in the second case I know the second cat, in the first I do not. The hypothesis under consideration is this: if I know E, then my knowing E constitutes an increase in coherence, in contrast with my not knowing E.

What is essential here is a *weltanschauung*, a conception of the universe. I have a more coherent conception of the universe in the second case cited than in the first. In the first case I am confronted with all sorts of questions that I cannot answer: where did that cat come from? How did he get in to the garden? Is he aggressive? And so forth.

But now consider what may seem to be a more difficult kind of case: there is a cat, Max by name, who lives in Kansas. George knows Max, Josef does not: ignoring all other factors, or "assuming them to be equal", does George have a more coherent conception of the universe than Josef? Though the difference is perhaps minute, I think one is forced to admit that George does have the more coherent conception. For what if someone were to attempt the laudable endeavour of a universal cat census? George would be able to evaluate the correctness of such a census in a way that Josef would not.

138. To jump to the case in which E is how to swim, what is then in question is having learned or not having learned to do something. Is having learned to do something more coherent than not having learned to do something?

If one focuses on that which is learned, then one must, I think, conclude that there is no reason to suppose that that which is learned is more coherent than that which is not learned. I have learned to swim; I have not learned to breathe (here ignoring special cases, such as when playing tennis, engaging in gymnastics and so forth): my swimming is certainly a less coherent process or activity than my breathing. My breathing is a genetically determined beautifully organized and regulated process: my swimming is pathetic. Nonetheless, I know how to swim (poorly) and it is not the case that I know how to breathe.

To focus exclusively on that which is learned would here be a mistake. For what is in question is the process of learning to do something. To focus on that which is learned is to adopt a synchronic stance. But, as we have already seen, such a stance is more appropriate in connection with knowing that p than in connection with knowing whether, or how, or where and so forth. Not that which is learned, but learning that which is learned is what is relevant here.

Does learning to do something increase coherence more than not learning to do something? The answer is clearly yes. For the process of learning to do something is a process in which matters are unified, organized and coordinated: learning to swim requires acquiring the ability to coordinate a complex sequence of movements.

Knowing George constitutes an increase in coherence where what is in question is a *weltanschauung*, a conception of the universe. But knowing how to swim constitutes an increase in coherence where what is in question is not a *weltanschauung*, not a conception of the universe, but rather a capacity to interact with that universe. (Here one can see how pragmatical an air a use of 'know how to' can lead one to breathe.) That the increase in coherence constituted by knowing George is of one kind, but that constituted by knowing how to swim is of another, in no way serves to establish that the meaning of 'know' shifts from case to case when one speaks of "knowing *E*". If one says 'That is a dangerous animal', the meaning of 'dangerous' does not shift from case to case when the referent of 'that' shifts from a tiger to a scorpion: in each case there is a genuine danger, but a danger of a different kind: one need not fear the tiger's tail.

139. The hypothesis in question is that if I know *E*, then my knowing *E* must constitute an increase in coherence. The hypothesis readily accomodates all cases considered. But that may suggest that it is too accomodating, and hence, not genuinely explanatory.

Suppose someone says 'I know the number 289': what can one make of such a remark? If we are to understand it, on the hypothesis in question, we must find some aspect of the matter that admits of considerations of coherence and of increasing coherence. The contrasts of truth and falsity, of existence and nonexistence, are not available. (If the number "exists", there is no question of its not existing.) And although the process of learning to do something is a process in which matters are unified, coordinated and so forth, it is not clear that if one learns the number 289, there is any process of doing that which moves in the direction of greater coherence.

I am not for a moment suggesting that it is senseless to say 'I know the number 289': I am saying that such a remark is not often made, and that, if made, would be likely to puzzle many. But if one embeds the remark in an appropriate discourse, the puzzlement can easily be dispelled: it is dispelled by invoking considerations of coherence.

'I know the number 289: it is the square of the seventh prime and it is the successor of 169 and the predecessor of 361 in the series of prime squares': embedding the remark in such a discourse serves to make clear at once that considerations of coherence are relevant; one identifies a set of structural features of the number.

Here one should not be tempted to suppose that 'I know the number 289' is simply a way of saying 'I know various truths about the number 289', for one could know various, indeed infinitely many, truths about a number and yet not know the number. Thus if one knows that *pi* is not a positive integer and is a number useful in trigonometric computations, one perhaps knows infinitely many truths about *pi*, that it is not equal to 1, to 2, ..., and yet one need not know the number.

'How well do you know *pi*? – To the twentieth place'. How well one knows a number depends, not on the quantity of truths known, but on their quality. One who knows simply that 289 is a positive integer, but nothing much else about it, could be said to know the number, but one would not say of him 'He knows the number very well' or 'He truly knows the number'. Such a person has only a limited knowledge of the number. In contrast, one who knows truths about the number that serve to indicate the achievement of a greater coherence could be said to know the number well: the quality, and not the quantity, of truths is what matters, and the qualitative difference is to be found in indications of coherence. Thus one who knows that 289 is the square of the seventh prime knows the number better than one who merely knows that 289 is a positive integer.

The mathematician, G. H. Hardy, wrote about Srinivasa Ramanujan:

He could remember the idiosyncrasies of numbers in an almost uncanny way. It was Mr. Littlewood (I believe) who remarked that 'every positive integer was one of his personal friends.' I remember once going to see him when he was lying ill at Putney.

I had ridden in taxi-cab No. 1729, and remarked that the number seemed to me rather a dull one, and that I hoped it was not an unfavourable omen. 'No,' he replied, 'it is a very interesting number; it is the smallest number expressible as a sum of two cubes in two different ways.'[56]

140. And now we are at a point from which one can see, and clearly, why the problems posed by skepticism cannot rationally be resolved by a simple act of renunciation.

Claims to knowledge are often suspect; skeptics, throughout recorded history, have scoffed at even the possibility of knowledge. There has always been, and still is, a simple way to still all such doubts: one could abandon any conception of knowledge; one could cease to use the words 'know' and 'knowledge', or any terms that could serve to convey the same conception. But in so far as we are rational beings, that is something we cannot do: it would be tantamount to committing intellectual suicide.

If one knows something, one's knowing it exemplifies and consti-tutues an increase in coherence. To seek to know is to seek a greater coherence: to know is to achieve it. To give up a use of 'know' and 'knowledge' (and any equivalents) would be to abandon any means of expressing the achievement. But a rational being would abandon the means of expressing such an achievement only if he abandoned the achievement: no rational being can do that and remain a rational being. (Why couldn't one abandon 'knowing' and 'knowledge' and speak instead of "believing truly" and "true beliefs"? If I have the true belief that p, that, too, must constitute an increase in coherence. But, in so far as knowing that p requires the satisfaction of conditions not requisite for believing truly that p, knowing that p must constitute a greater increase in coherence than merely believing truly that p.)

One seeks to know, not for the sake of imparting that knowledge to others, though, on occasion, that may be a factor, but for the sake of achieving a greater coherence: for a rational being, the value of coherence, and hence of knowledge, is intrinsic; it is of value in and of itself. The price of life is death, and the consequent loss of all the values achieved during life that were of value only so long as one lived. But the value of knowledge is transcendent.[57]

To know that the price of life is death (or that $2 + 2 = 4$, or how to play tennis, or no matter what) is, perhaps, not much, but it is at least partial compensation for the ordeal. That the achievement is an achievement of the moment does not matter. What is then done cannot then be undone. Which is why skepticism is a worry: for what if one had not achieved what one thought one had? This is a worry we shall turn to later.

Here one can see that knowledge, like life itself, is at odds with the fundamental metaphysical truth expressed in the fall of Humpty Dumpty: not all the king's horses and all the king's men could put Humpty Dumpty together again. Entropy moves in the direction of incoherence: knowledge moves contrariwise. (Which makes one wonder about those who admire the course of the natural: what is more natural than to die? Despite the claim to the contrary, that is not a consummation devoutly to be wished for.) Knowledge is an intrinsically valuable temporary achievement, an increase in coherence in our limited world.

141. (It is perhaps tempting, in the light of their similarities, to attempt to identify the acquisition of knowledge with a decrease in entropy, with negentropy: but that is, I think, a temptation that one must resist. Entropy is an abstract physical conception associated with the Second Law of Thermodynamics; whereas coherence is essentially a matter of logic: any connection between the two is, if not fanciful, at best, remote.

This is not to overlook the fact that knowledge is acquired by human beings, and human beings are living organisms in a physical world governed by physical laws. That the acquisition of knowledge is essentially a psychical phenomenon is not to be denied; but neither is the fact that psychical phenomena require physical realizations. Nor can one overlook the fact that irreversible processes are involved in the acquisition of knowledge, in the achievement of an increase in coherence. As Niels Bohr has said:

From a biological point of view, we can only interpret the characteristics of psychical phenomena by concluding that every conscious experience corresponds to a residual impression in the organism, amounting to an irreversible recording in the nervous system ...[58].

The acquisition of knowledge implicates learning, and learning implies an alteration in neurophysiological structures. But that such alteration corresponds to negentropy is another matter.

If you put a man into a Humpty Dumpty plight by scrambling his brains, you wipe out whatever knowledge he had: but you also wipe out whatever misinformation he had. The acquisition of mis-information also implies an alteration in neurophysiological struc-tures: there is no good reason to suppose that the acquisition of knowledge constitutes a decrease in entropy, while the acquisition of misinformation constitutes an increase in entropy. Yet the acquisition of misinformation is a move in the direction of incoherence, while the acquisition of knowledge is a move in the direction of coherence.

Fancifully speaking, one may say that knowledge is power, and power is energy, and so the acquisition of knowledge implies negentropy. But though this may be a pleasing fancy, it is no more than that. Not the acquisition of knowledge, but the utilization of knowledge can, as Szilard has shown, be associated with negentropy.

In Szilard's discussion of the Maxwell demon problem, the demon was regarded as "receiving information" about the particle motions of a gas, this information enabling him to operate a heat engine and set up a perpetuum mobile; the demon was making use of his information, not simply receiving it and passing it into storage.[59]

But one can have knowledge without acting on the knowledge in any way.

Perhaps the strongest somewhat plausible claim that one could make here is that knowledge constitutes a potential for negentropy, and that the acquisition of knowledge constitutes an increase in that potential.)

142. (Having disclaimed any attempted identification of knowledge with negentropy, it may help to add another disclaimer. Truth is, on anyone's account, a difficult conception, but, as far as I can see, it has nothing whatever to do with coherence.

'Truth' is the nominal corresponding to the adjective 'true'.
'Knowledge' is the nominal corresponding to the verb 'know'. It is
difficult to compare 'truth' and 'knowledge' because it is difficult to
compare 'true' and 'know', an adjective and a verb. Even so, compare
the following two statements: 'If he knew that 16661 is a prime, it
does not follow that he knows that 16661 is a prime'; 'If it was true
that 16661 is a prime, it does not follow that 16661 is a prime'. The
first statement is true, the second false. The difference is, of course,
that knowledge, but not truth, may erode in time. If I state 'I knew
that p' that implies that I know that p; whereas if I state 'It was true
that p', that does not imply either that it is true that p or that I think
that it is. (If anything, if I state 'It was true that p', the suggestion is
that, perhaps, it is no longer true, though the remark might, instead,
be a concession.) Truths do not accrete, neither do they grow. There
is no extant corpus of truths in the way there is an extant corpus of
knowledge. If one were to wipe out all automata (both organic and
mechanic), one would have wiped out all knowledge: there is no way
of eliminating the truth.

If one takes seriously the actual use of the words 'true', 'truth', 'truly'
and so forth, the conception that is evidently involved is not that of
coherence, but rather one of conformance. (I hasten to add that
conformance is not "correspondence": 'He conforms to the rules'
makes sense; 'He corresponds to the rules' does not (literally).)

A true friend is one who conforms to the ideals of friendship. A true
description is one that conforms to the requirements imposed by that
which is to be described. To true up the lines of a tennis court is to
make them conform to the standards set in the rules of play. To be
true to one's spouse is to conform to the sexual limitations enjoined
by a marriage contract. To see things in a true light is to see things
in such a light that the way they look conforms to the way they are:
it is not to see them in the light of a laser, a truly coherent light.

Although the coherence of a theory may be conspicuous, one does not
ordinarily speak of a theory as "true": theories are, rather, correct,
sound, well-founded and so forth.

But these few remarks shall here have to suffice: I am not here concerned to articulate a theory of truth, only to disclaim any commitment to one such theory.)

143. One must be on guard against the temptations of knowledge. Knowledge is intrinsically valuable, but pain, misery and anguish are intrinsically disvaluable.

Kant was seduced by the lure of knowledge: it is not always wrong to lie. To lie may be to increase local coherence at the expense of overall coherence. But the intrinsic values achieved may more than compensate for the decrease in overall coherence. (It's a pity that Kant never pondered the plight of Humpty Dumpty.) And such monstrous obscene experiments as Harlow's "Well of Loneliness" also betray an incredible insensitivity to the place of knowledge in a rational scheme of values.[60]

Kafka said: in a war between the world and me, bet on the world. Knowledge is a momentary achievement. It, as everything, is subject to inevitable entropic degradation. But so long as time exists, the past is fixed fast in time like glass, and cannot be undone. ("En ceste foy je vueil vivre et mourir".) But a decrease in suffering is also an intrinsically valuable achievement. And even if one views matters *sub speciae aeternitatis*, or if one doesn't, the balance sheet of trivial knowledge versus real misery will remain the same.

As Peirce perhaps realized (for he was prompted to say that logic is based on ethics and ethics on aesthetics), the intrinsic value of knowledge is proportional to the extent that it has a bearing on human affairs. This is not to say that the intrinsic value of knowledge is somehow fundamentally extrinsic. All truths are undoubtedly truths, but that does not mean that all truths are on a par: some are significant and some are not. Any piece of knowledge has intrinsic value, but some are more valuable than others: for though man is not the measure of all things, his interests are the only real measure for the intrinsic value of knowledge.

144. If I know that p, then p must be true. That p must be true is explicable in terms of the requirement that my knowing that p must constitute an increase in coherence. But if I know that p, not only must p be true, but I must be in a position in which the possibility of my being in error with respect to whether or not p may safely be discounted. That, too, wants explaining. What, for example, does the possibility of error have to do with my knowing George, or my knowing how to swim?

The answer is, of course, nothing at all. But there's no reason to think it would or should. Figuratively speaking, p invites ignorance and error; George asks for privacy and lives apart; while swimming prepares the way for an inability and drowning.

One must proceed step by step: if I am to know that p, p must be true. But there are no end of truths that I am ignorant of. If I am to know George, then he must exist. But there are no end of people who are strangers to me. If I am to know how to swim, there must be such an ability as the ability to swim. But there are no end of abilities that I do not have or concern myself with.

If I am to know E, I must be in a position to know E; so far, so good, but somewhat circular, which may alarm some. But one can step outside this circle once E is specified. If I am to be in a position to know that p, I must be in a position which facilitates the avoidance of ignorance with respect to p, and of error with respect to whether or not p. I am in a position to know George if I am in a position which facilitates overcoming his privacy and encountering him. I am in a position to know how to swim if I am in a position which facilitates concern with the ability and success in acquiring it.

CHAPTER XII

SKEPTICISM

145. I KNOW that *p* if and only if *p* is true, and I am in a position in which any possibility of my being in error with respect to *p* may safely be discounted. If a skeptic claims that there is no position which I can be in, in which there is no possibility of error, I agree. No one is, or ever is, infallible. But serious skeptical problems are posed by the claim that no one can ever be in a position in which the possibility of error may safely be discounted, thus by claims that, for any given *p*, there is no position that is safe with respect to *p*. Or, more simply, for any given *p*, no one ever knows, or ever can know, whether or not *p*.

Skepticism is a plague, both to the faithful,[61] and to the faithless, such as reasonable epistemologists. But there is a portentous truth in skepticism that has, as far as I know, been overlooked. Skepticism is the easy way, for altogether explicable reasons. It is no accident that one can easily lead students to adopt a skeptical stance, that bright eyed bushy tailed college freshmen, and innocents innocent of epistemological theory, when confronted with epistemological problems, are at once inclined to adopt a skeptical stance.

For it is possible to prove that someone does not know something in a way that it is not ever possible to prove that someone does know something. One can, of course, reason that no one can know that *p* if *p* is a contradiction. But that is not here to the point. For if *p* is a contradiction, then *p* is not true. The existence of contradictions lends no support to skepticism. What is wanted here is a case in which *p* is true, and, to make matters better, a case in which someone claims that *p* is true.

Given that George claims that *p*, and that *p* is true, is it possible to prove that George does not know that *p*? What is in question here is a proof, not hand waving. Given the exigencies of exposition, I have, I confess, resorted to appropriate hand waving on occasion. The order of exposition is not the order of proof.[62] So I claimed that Josef didn't in fact know that the horse Josef K would win the race even though he did, Josef K that is, win. I said it was just a matter of luck that Josef was right, and knowledge isn't just a matter of luck. That's true enough. But I didn't prove it. Given a correct analysis of knowledge, one can prove it.

146. Let s be that the next spin of the roulette wheel will end up red. Ken claims to know that s, whereas Don claims to know that not s. So one of them is bound to be in error (assuming, of course, that there will be a next spin of the wheel and so forth).

Let us say that Ken is in position Sk, Don is in position Sd; now let me characterize the positions that they are in. Ken and Don are identical twins, and have received the same lack of education. Each are cowpokes living on a ranch in Idaho. Each has decided to take, and has taken, a brief vacation. Ken took off for Reno, Don for Las Vegas. In Reno, Ken played roulette, and lost, but witnessed and remembered a sequence of the roulette wheel he played that was constituted by three blacks, two reds, one black, four reds and then black. In Vegas, Don played roulette, and lost, but witnessed and remembered a sequence of the roulette wheel he played which was constituted by three blacks, two reds, one black, four reds and then another red. Don and Ken are now in Carson City, playing roulette again: they witness three blacks, two reds, one black, and then four reds. At that moment, Don, placing his last chip on red, claims to know that the next spin will turn up red, while Ken, making a side bet and betting his last chip, claims to know that the next spin will not turn up red.

One of them is going to be right. But neither one knows what he claims to know.

147. Don claims to know s, Ken to know that not s. Don is in position

Sd, Ken in *Sk*. Both *Sd* and *Sk* have been characterized in such a way that there is, and can be, no way in which *Sd* is safer than *Sk* with respect to *s*. (If I have forgotten or overlooked something, then overlook my oversight and make the appropriate alterations.) For it is essential here that *Sd* equal *Sk* with respect to safety. That is to say, one who is in position *Sd* is as safe, and no safer, than one who is in position *Sk* with respect to *s*.

If so, it follows, more surely than the night the day, that neither is in a safe position with respect to *s*, and, hence, on the presented analysis, neither knows what he claims to know, even though one or the other is bound to be right (given that there is a next spin of the wheel and so forth). For if in position *Sd* one claims that *s*, but *s* is not true, then *Sd* cannot be a safe position with respect to *s*. And the same is true of *Sk*. But since *Sd* equals *Sk* in safety, neither can be safe, and hence neither Don nor Ken knows what he claims to know. I said that it is possible to prove that someone doesn't know something in a way that it is not ever possible to prove that someone does know something. But if I prove that Ken does not know that *s*, haven't I proven that I know that *q*, where *q* is that Ken does not know that *s*?

Hardly: for one should reflect on the fact that it is quite possible to prove theorems and yet not understand what one has proved. Furthermore, one may prove something and not even realize that one has proved it. One may mistakenly suppose that one's proof is fallacious.

In the preceding example I did prove that Ken does not know that *s*. And I am quite prepared to claim that I know that, in such a case, Ken does not know that *s*. But I have not tried to prove that I know that, although I have tried to show that I know that by presenting the proof that Ken does not know that *s*.

148. What is of moment here is the fact that the proof presented above cannot be reversed: one cannot ever prove that someone knows that *p* in the way one can prove that someone does not know that *p*. That

one cannot do so is owing to the difference between positive and negative existential statements.

There is an absolutely fundamental difference between positive and negative statements that surfaces only when one is concerned with indefinite or infinite domains. If one is, as a rat, placed in a T-maze, the instruction 'Go left to find your food!' is wholly equivalent to 'Don't go right to find your food!': given a finite set of alternatives, the difference between positive and negative can vanish. But if one were engaged in the (admittedly curious) process of writing down numerals at random, there is no positive equivalent for the instruction 'Don't write the numeral 17!'.

Negative existential statements, such as 'There are no arachnids in the Vatican', have a theoretic depth not to be found in connection with positive existential statements, such as 'There is at least one arachnid in the Vatican'. Both the positive and the negative statement have a theoretic aspect, for if one is to understand either statement, one must have some conception of what an arachnid is. But there is a vast difference in the theoretic depth of the two statements.

One could establish that there is an arachnid in the Vatican by finding one, say, in a crevice somewhere. One would, of course, have to consider whether what one had found was indeed an arachnid, and not, say, an odd beetle. But one could not establish that there was no arachnid in the Vatican by failing to find one. Furthermore, if one has found an arachnid, the size of the Vatican becomes irrelevant; but if one has not found one, the relevance of the size of the Vatican looms even larger.

149. In the preceding example, it may have been noticed that there was a reciprocity between the size of what was in question, namely an arachnid, and the area of investigation, the Vatican. If one increased the size of the object sought for, say, by switching from an arachnid to an elephant, and decreased the area of investigation, say, from the Vatican to a small motel room, the theoretic difficulty in establishing that there is an elephant in the room would be on a par

with the theoretic difficulty in establishing that there is no elephant in the room: neither would pose any considerable problem.

This is no accident: it exemplifies an aspect of logical structure that is often unnoticed. A failure to appreciate this aspect of logical structure is to be found in connection with the so-called "fallacy of the sorites".

Consider the argument: a man with only one penny to his name is a poor man. If a man is poor and you give him one penny, he still is poor. Therefore a man with two pennies is poor. This is a perfectly valid argument. But it displays a remarkable aspect: it is subject to degradation in time. For if one could (*per impossible*) repeat the process a billion billion times, one would arrive at the patently false conclusion that a man with a billion billion pennies was poor.

(Here one should not suppose that the problem is created by employing such a term as 'poor', a term that admits of degrees. For the same problem is posed by switching to such an example as: something that is a table remains a table if its legs are shortened by 1/100th of an inch. If one repeated the process over and over, one could arrive at the curious conclusion that a slab of wood on the floor was a table: at best, it could only be a table top.)

To put the matter succinctly, the argument about the man with only one penny is a perfectly fine argument if it is not overused, for owing to its temporal cast, it is subject to temporal degradation. How often one can employ such an argument depends precisely on the character of the increment (or decrement) and the particular predicative attribution at issue. If one increases the increment, there may have to be a corresponding decrease in repetitions. There may have to be, but there need not be; for one could alter the character of the domain: consider being, not a poor man, but a man of limited means, or better, not being a man of umlimited means; then regardless of the size of the increment and the number of repetitions, it will remain true that the man is not a man of unlimited means. Alternatively, leaving the domain unchanged, if one adjusts the increment appropriately, one can go on forever (in theory at least). For consider: a man with only two pennies, one of which has been given to him, is a poor man. If one gives a poor man 1/2 of what he has last been given, he remains poor. Therefore a man with 2 and 1/2 pennies is poor; therefore a man with 2 and 3/4 pennies is poor; and so on.

The temporal degradation displayed in the "fallacy of the sorites" is not peculiar to that argument form. It is a feature of all simple inductive processes in the actual world. Thus suppose I take one step in a dark room and touch no wall with an outstretched hand; then another step and again nothing: should I conclude that if I take still another step, I will again touch no wall? That obviously depends on the size of the room, my location in it, and the pattern of my movements: if am appropriately placed in an enormous room, and I take very small steps, I may go on for a long time; whereas if I am walking in a small circle in a tiny room, I can go on till I drop. Simple inductions are simple nonsense if one does not take into account the character and extent of the domain of attribution and the character of the attributions in that domain.

The need to take these matters into account is not peculiar to arguments cast in a temporal mold: it is simply more obvious in such cases. But even valid atemporal syllogistic forms are subject to the same requirements. All Romans admire Rome; Caesar is a Roman: hence Caesar admires Rome. True, but: All Romans turn away from Caesar; Caesar is a Roman: hence Caesar turns away from Caesar? Here considerations of coherence require the scope of the quantifier to exclude Caesar. But what about: All Romans kick Caesar; Caesar is a Roman: hence Caesar kicks Caesar? Perhaps Caesar kicks himself, perhaps not.

150. The difference in theoretic depth between positive and negative existential statements is directly attributable to the different ways in which positive and negative statements determine the character and extent of the domain of attributions. Perhaps the clearest example of this is to be found in the so-called "paradox of confirmation".

Form the hypothesis that all crows are black; if so, by the principle of the contrapositive, one can infer that all nonblack things are noncrows. Hence a piece of white chalk would seem to contribute to the confirmation of the original hypothesis. Of course it doesn't. Why not? What is the relevant domain if one says 'All crows are black'? At least a domain that includes all crows. What is the relevant domain if one says 'All nonblack things are noncrows'? At least a domain that includes all nonblack things. But such a domain is irrelevant in connection with 'All crows are black'. What is wrong is simply that the principle of the contrapositive has been misapplied.

For consider a restricted domain, say, a cage occupied by twenty animals. We are emptying the cage at random, one animal at a time. We have so far removed two black crows and three white storks. With respect to this domain, we form the hypothesis that all crows are black, and infer that all nonblack things are noncrows. We then remove a white rabbit. Does the removal of the white rabbit contribute to the confirmation of our hypothesis? Yes, for prior to the removal of the white rabbit, there were 15 animals left in the cage. Hence we had 15 chances left to be proven wrong, to find a nonblack crow. Once we have removed the white rabbit, we have only 14 chances left. So things are looking up for our hypothesis.

It is worth noting that one could proceed in another way: for one could alter the character of the domain by extending it, but compensate for the extension by altering the character of the attributions in the domain. Thus the principle of the contrapositive poses no problem if one says 'All crows are black birds' and one assumes a spatio-temporally finite domain of birds. One can then infer that all nonblack birds are noncrows, and finding a white stork will then contribute, albeit minimally, to the confirmation of our hypothesis. (And one could, of course, go the other way and make things worse by invoking an infinite domain and allowing 'nonblack' to be used to characterize anything such that it is not the case that it is black, and the same for 'noncrow'. In that case, that all nonblack entities are noncrows might seem to be given some confirmation by the number 17, for it is not the case that it is black and not the case that it is a crow: but given the infinite domain, one is no better off than before.)

151. The impossibility of proving that someone knows that p in the way that one can prove that someone does not know that p, arises from the temporal character of knowledge.

It is predictable, given the (correct) assumption that knowledge is constituted by an increase in coherence, and, so, analogous to a decrease in entropy. A decrease in entropy is, at best, a local achievement within the limited horizons of an unlimited universe. But knowledge has an indefinitely large domain. Explicit attributions of knowledge can span millenia: the ancient Hindus knew that a language has phonological, morphological and syntactic structures. In saying that they knew that, I indicate, not that I knew, but that I know that too. Such knowledge may be lost in the interim, for that Plato knew that is unlikely, given the remarks in the "Theatetus". But if the Hindus knew that p, no one can ever say truly 'I know that not p'. Given such an indefinite domain, a proof that one knows that p is an impossibility.

To prove that one knows that p, assuming that one is in position Si, one would have to prove a negative existential statement in an indefinite domain: one would have to prove that there never was, nor is, nor ever will be a position Sj (where i is not equal to j) such that Sj is as safe, or safer, than Si with respect to p, and in which one could claim that not p.

152. One should not be overimpressed, and certainly not overwhelmed, by the impossibility of a proof that one knows that p.

Proofs aren't all that impressive. Many obvious truths do not admit of proof, and any proof has its limitations. One cannot, in fact, prove that $2 + 2 = 4$: one can give a mock proof, but nothing more. An elementary proof takes the form of a series of premises, $P1 \ldots Pn$, and a conclusion, C, thus $P1 \ldots Pn/C$. One can provide an argument of the form $P1 ... Pn/(2 + 2 = 4)$, and one can call it a "proof", but it proves nothing. For consider what one would say of an argument of the form $P1 ... Pn/(2 + 2 = 5)$: obviously it would be a *reductio ad absurdum*, a proof that something was wrong either with the premises or with the principles of inference employed. To provide a proof that C, C must be in need of a foundation which is supplied by the proof. The premises must serve to increase the plausibility of the conclusion. But if so, an argument of the form $P1 ... Pn/(2 + 2 = 4)$ is turned the wrong way round, for, at best, the conclusion might tend to increase the plausibility of the premises and the principles employed.

That one cannot have a proof that one knows that p does not confirm skepticism: it tells us something of the character of knowledge, that its domain is indefinitely spatio-temporally extended. It also tells us something already known: that our methods of proof have their limitations. To attempt to derive a skeptical conclusion from the famous Godel incompleteness theorem would be a gross error arising from a confusion between, or an attempt to conflate, the conception of knowledge and the present conception of proof. If knowledge and proof coincided, one could not know the truth of the axioms of an axiomatized theory: for if p is an axiom, the mock proof that p from the axiom p would not warrant a claim to knowledge: genuine proof is a means of extending knowledge; it does not, and cannot, constitute knowledge. (The legitimacy of various means of extending mathematical knowledge is the issue dividing Intuitionists, Constructivists and Classical mathematicians.)

153. Even though one can't prove that one knows that p, one can prove that one knows how to swim, and one can prove that one knows George.

One can prove that one knows how to swim by successfully performing the appropriate actions, namely by swimming an appreciable distance. Analogously, one could prove that one knows George by singling him out in a crowd, greeting and being greeted by him and so forth. This disparity wants explaining, though perhaps by now the reasons for it are reasonably clear.

Knowing how to swim and knowing George have a positive existential import in contrast with the negative existential import of knowing that p. Knowing that p requires the avoidance of ignorance and error, whereas knowing George requires overcoming his privacy and separateness by a positive encounter, and knowing how to swim requires overcoming an inability and possible failure by a positive acquisition. Given the positive existential import of knowing how to swim and knowing George, proof becomes possible.

But even though one can't prove that one knows that p, one can, as axiomatic systems show, prove that one knows that q, on the assumption that one knows that p and one knows that p entails q. Such proofs may be called "relative proofs". A relative proof may constitute a genuine extension of knowledge. Furthermore, relative proofs are both readily available and of considerable importance. For such proofs serve to establish, at the very least, an increase in local coherence: one can't rationally ask for more.

CHAPTER XIII

A SAFE POSITION

154. A SAFE POSITION is one in which the possibility of error may safely be discounted. But when is one in such a position? To answer this question, we must attend to positions in some detail.

155. Characterizing a position is no simple matter. I mean to go slowly about it.

I shall use the expressions 'Ki', 'Kj' and so forth, as dummies for the names of persons, '$\{si\}$', '$\{sj\}$' and so forth for sets of conditions, and 'Si', 'Sj' and so forth for positions.

Not just any set of conditions can constitute a position. To do so, some members of the set must have reference to persons (or automata, organic or otherwise). Thus if set $\{si\}$ includes only the condition that it be winter, then $\{si\}$ does not constitute a position, hence there is no associated position Si. But if set $\{si\}$ also includes the condition of being a person with a cold, then $\{si\}$ constitutes the position Si, a position that someone can be in.

If set $\{si\}$ is so specified, then anyone who has a cold when it is winter is in position Si. But it should be noted that if Ki is in position Si, it does not follow, and it would here be incorrect to say, that Ki satisfies the conditions of $\{si\}$. Ki does satisfy one condition of $\{si\}$, namely that of being a person with a cold. But Ki does not, and cannot, satisfy the condition that it be winter: only a season can. (This is not to deny that one could characterize the set of conditions, $\{si\}$, in another way, say, by replacing the condition that it be winter with the condition of being a person when it is winter. But I shall not adopt such a way of speaking.)

We can then say: Ki is in position Si if and only if the associated set of conditions, $\{si\}$, are satisfied with reference to Ki.

156. It is possible to specify the set of conditions $\{si\}$ in such a way that Ki, and only Ki, is always in position Si, or never in it, or sometimes in it and sometimes not, or in such a way that Ki and others may or may not be in it, and so forth.

Thus if $\{si\}$ includes only the condition of being Ki, then Ki is always in Si. If $\{si\}$ includes the condition of being Ki and certain other conditions as well, then Ki may or may not be in Si, but no one else can be. For example, only George can be in the position of having legal grounds to divorce George's present (nonbigamous) wife. Whether or not George is in that position depends on further conditions. If $\{si\}$ includes the condition of not being Ki, then Ki is never in position Si. Thus George is never in the position of being able to meet himself on the street. (If these observations seem hardly worth making, wholly obvious, it is worth calling to mind such remarks as "Only the person in pain really knows whether or not he is in pain".)

157. One must attend to the difference between a set of conditions that constitutes an occupiable position and a set that either constitutes an unoccupiable position or simply fails to constitute a position. There are various possibilities, depending on the particular set in question.

If set $\{si\}$ includes only the condition of being a citizen of the United States, then Si is an occupiable position; whereas if it also includes the condition of having spoken with Genghis Kahn, Si is an unoccupiable position: that it is unoccupiable is owing to the fact that it is physically impossible to satisfy the conditions requisite for being in Si. But if it were logically impossible to satisfy the conditions of $\{si\}$, then $\{si\}$ could not constitute a position. Thus if $\{si\}$ includes the condition of having proven Fermat's last theorem, then Si is, at present, unoccupied and, conceivably, may not even be a position: there's no way of knowing at present.

Ki is in position Si if and only if the set of conditions, $\{si\}$, is satisfied with reference to Ki. Ki knows that p only if Ki is in Si, and Si is safe with respect to p, thus if and only if the set of conditions, $\{si\}$ is satisfied with reference to Ki, and $\{si\}$ constitutes a safe position with respect to p. The problem then is to find a nontrivial characterization of the set of conditions, $\{si\}$, that can serve to indicate whether or not it constitutes a safe position.

A trivial characterization of a safe position, Si, could be supplied by having the set of conditions, $\{si\}$, that serves to constitute position Si, include the condition that p be the case. For, obviously, if Ki were then in Si, it would follow logically that p is the case. But such a tautological characterization would here be pointless. Hence, let it be understood that $\{si\}$ does not include the condition that p be the case.

158. It is essential to appreciate the fact that it is not possible to provide a specific, nontrivial characterization of the set of conditions, $\{si\}$, that will serve to establish that $\{si\}$ constitutes a safe position. By "a specific characterization", I mean a characterization by an enumeration of the members of the set; and by a "nontrivial characterization", I mean a characterization that neither includes nor entails the condition that p be the case.

Consider a case in which I know that I am not guilty of a certain crime; I am in position Si, and in Si, I know that p, thus $\{si\}$ constitutes a safe position. The prosecuting attorney claims that not p. He has a case against me based on circumstantial evidence, none of which I can explain. My fingerprints are on the weapon, but I do not know how that happened; the victim was seen entering my house, but I did not know that the house had been entered, nor do I know why; the victim had left a note saying that I had asked him to enter the house, though, in fact, I had not; I had both motive and opportunity; and so forth. Everyone believes that I am guilty, but I am not. I, and I alone, know that.

If I were asked, though the question would be curious, how I know that I am not guilty, I might reply that I recall being in my study at the time that the crime was committed elsewhere. But if that is true, then either that entails that I am innocent or it fails to establish that

I am in a safe position. If I were then asked how I know that my recollection is correct, there is nothing I could say that could serve to establish that it was; I could, of course, say things to the effect that I do not suffer from a defective memory, am not subject to lapses, and so forth. But, again, my answers would either entail that I am innocent or they would fail to establish that I am in a safe position.

159. Since no specific nontrivial characterization of the position I am in will serve to establish that it is, indeed, a safe position, one must focus on, not the membership of the relevant set of conditions, but, rather, on characteristics of the set itself. But that, too, poses something of a problem.

I am in position, Si, constituted by $\{si\}$. Let $\{sj\}$ be the set of conditions that serves to constitute the position of the prosecuting attorney, in which he claims that not p. Let $\{p\}$ and $\{-p\}$ be the sets of conditions constituted by p's being, and not being, the case, respectively. Then, given that knowledge constitutes an increase in coherence, it must be the case that the union of $\{si\}$ and $\{p\}$ constitutes a greater increase in coherence than the union of $\{sj\}$ and $\{-p\}$. If, however, one attends simply to the matter of local coherence, there is no warrant for such a conclusion. Obviously, given the circumstantial evidence, the prosecuting attorney has the more plausible case: even though it would be an error, and a miscarriage of justice, any reasonable judge would have to conclude that I am guilty. And that is simply owing to the fact that the local coherence of the union of $\{sj\}$ and $\{-p\}$ is appreciably greater than the local coherence of the union of $\{si\}$ and $\{p\}$.

160. Here one must confront an uncomfortable, but unavoidable, counterfactual: if the truth were known, one would know that p is true and that, when I claimed to know that p, I did indeed know it; position $\{si\}$ was, in fact, a safe position.

What is wanted here is a more global perspective from which one can discern the difference between a genuine increase in coherence and a mere local increase at the expense of an overall decrease. Let $\{O\}$ and $\{C\}$ be sets of conditions constituting open and closed systems, respectively. Then, given that the prosecuting attorney has the more plausible case, one can admit that the union of $\{C\}$, $\{sj\}$ and $\{-p\}$ may be more coherent than the union of $\{C\}$, $\{si\}$ and $\{p\}$; but given that I know that I am innocent, the union of $\{O\}$, $\{si\}$ and $\{p\}$ must be more coherent than the union of $\{O\}$, $\{sj\}$ and $\{-p\}$.

That a nontrivial characterization of a safe position necessitates an appeal to an open system of knowledge should not be surprising. It is a reflection of the fact that knowledge has an indefinite domain. Knowledge is cumulative and evolutionary. It admits of both accretion and genuine growth: the knowledge one has already acquired may enable one to acquire further knowledge. But the emergent system of knowledge, constituted by the set of conditions $\{O\}$, is also cautionary and restrictive: if someone once knew, or someone will know, that p, then no one can ever know that not p.

Let Si, Sj, ..., be occupied or occupiable positions. If Ki, in Si, is prepared to claim that p, whereas Kj, in Sj, is prepared to claim that not p, then Si and Sj are what may be called alternative conflicting positions. The only requirement for the safety of a position is that it be safer than any alternative conflicting position. One can then say: Si is safer than Sj if and only if the union of $\{O\}$, $\{si\}$ and $\{p\}$ is more coherent than the union of $\{O\}$, $\{sj\}$ and $\{-p\}$. This means that comparisons of relative safety will require attention to far ranging considerations.

161. There is no great theoretic problem, though there may be considerable factual problems, in determining the relative safety of two positions. For all such determinations are made on the basis of considerations of coherence, of completeness and consistency.

George and Josef want to know whether or not p is the case. George looks at the sky and Josef consults a deck of fortune telling cards. George claims to know that p, Josef claims to know that not p.

Suppose p is that the sky is gray. Then Josef is being weird. Suppose p is that the cards say that the sky is gray. Then George is being weird. Let us say that, having looked at the sky, George is in position Sg, and, having consulted the cards, Josef is in position Sj. Let p be that the sky is gray, and let q be that the cards say that the sky is gray. Then with respect to p, Sg is a safer position than Sj; whereas with respect to q, Sj is safer than Sg.

162. That with respect to p, Sg is safer than Sj is a fact. It is not logically impossible that it have been otherwise, at least in the sense that it is not self-contradictory to suppose so.

It is conceivable that actually looking directly at the sky might have had a serious effect on one, distorted one's vision, and color vision, in particular. (To find out how the sun looks, it's not the naked eye that is wanted.) And it is faintly and just conceivable that the character of the light served in some way to determine what one would read in the cards: the gray light filtering or sunlight streaming in the windows might have had some unsuspected influence on the cards, or more plausibly, on their readers. (Is this really conceivable? I doubt it, but it doesn't matter at the moment.)

Despite all this, Sg is safer than Sj with respect to p. But what if someone were to deny it? 'Then he would be wrong'. No doubt, but how and why is still to be explained.

163. George and Josef want to know whether the sky is gray. George looks at the sky and Josef consults his cards. George says that it is and Josef says that it isn't. They dispute the matter. George invites Josef to look at the sky. Josef invites George to look at the cards. George agrees that the cards say that the sky isn't gray. Josef admits that the sky looks gray to him then and there. Even so, they still disagree.

If we say that George is right, we then have the problem of reconciling the statement that the sky is gray with the statement that the cards say that the sky isn't gray. If we say that Josef is right, then we have the problem of reconciling the statement that the sky isn't gray with the statement that the sky looks gray to George and Josef then and there.

That with respect to p, Sg is safer than Sj does not depend simply on p, Sg and Sj. It is also necessary to consider q, that the cards say that the sky isn't gray, and r, that the sky looks gray to George and Josef then and there.

And that is not all. For whether we side with George or with Josef, there is still more to consider. If we side with George, then some explanation is wanted why the cards say what they say, why they are wrong. If we side with Josef, then some explanation is wanted why the sky isn't gray when it looks that way to both George and Josef. To explain why the cards say what they say, we shall have to consider how the assignment of values to cards is made, how cards are shuffled, dealt and so on. To explain why the sky isn't gray, when it looks that way to both George and Josef, we shall have to attend to questions of optics, physics, psychology and so forth. Inevitably, no matter whom we side with, the ripples of inquiry will spread out.

164. Despite the complications, it is, or should be, obvious that Josef is wrong: position Sj is less safe than Sg.

Josef's view is that a way to tell whether the sky is gray is to consult the cards, never mind looking at the sky. But Josef admits that the sky looks gray to him when he looks. Consequently, he has the problem of reconciling his claim that the sky isn't gray with his admission that the sky looks gray to him. The reconciliation to be effected requires attention both to matters of consistency and completeness. Here the appearance of consistency is relatively easy to achieve, since the compound statement, 'The sky isn't gray, and yet it looks gray to me at the moment' is evidently not self-contradictory. Whether or not one's account would still be consistent when further details have been filled in is, of course, another matter. However, I am inclined to suppose that the difficulty here is primarily one of completeness.

For Josef's view to be coherent, its various details must all fit together. Obvious gaps must be filled. But when one attempts to render Josef's view more coherent by completing it, by supplying the requisite details, the task becomes intolerably complex. In fact, one cannot do it.

If we say that a way to tell whether the sky is gray is by consulting the cards, then how do we account for the fact that an arbitrary assignment of values to cards nonetheless manages to correspond to the state of the weather? What if we shuffle the cards and read them again? We are likely to get a conflicting result. Even more, if the assignment of values is determinate, we can calculate exactly what the probability is of our getting a conflicting result. And still more, if we have two persons with two decks of cards, each giving an independent reading, we can calculate exactly what the probability is of their arriving at different answers. Suppose they do arrive at different answers: what shall we say? How shall we explain one answer away? In the case of two people looking at the sky, making opposing claims, we know what to do, where to look, what to say and so forth. Possibly something is wrong with the eyes of one of them, possibly one of them is color blind.

Here it may be objected that the difference between George's and Josef's views is that George's view has been worked out, Josef's hasn't: thus the conception of being color blind is already available to work out details in George's view. The requisite conceptions are not yet available for Josef's view.

But the reason why the requisite conceptions are not available is simply that they are completely arbitrary, *ad hoc*, contrived to fit the exigencies of Josef's position. To hold on to Josef's view, we must either suppose that there is a causal relation, or some type of functional correlation, between the configuration of the cards and the state of the weather, or else we must scrap many of our notions of causality. If we are to have some conception of a causal relation between the configuration of the cards and the state of the weather, then the conception must be articulated, or must, at least, be capable of articulation to some degree. What are some of the details? Does the color of the deck matter? Or the thickness of the cards? Is there a crucial distance between cards when they are laid out on the table? And so on. And if we manage to dream up answers to these questions, what sort of tests will be relevant? How shall we explain away the fact that when we perform obvious tests, the answers do not seem to fit with the view? And so on.

Though one may say, for the sake of argument, that it is conceivable that Josef's view be correct, strictly speaking, one can't really, fully, conceive of it. One can begin to conceive of it, but one can't really bring it off. At any rate, not this one, not today.

165. Given that it is possible to establish the relative safety of alternative positions, and given that one can establish the positive unsafety of a position by showing that it is no safer than an alternative conflicting position, it is then not difficult to see what is wrong with some familiar accounts of inductive reasoning.

Suppose that what is in question is whether the next object of a certain sort to be seen will be blue. Both George and Josef are in the position of having observed n objects of the sort in question, and every one was blue. Here it is essential that $2 \times n$ be a countable number, thus reasonably small, or at least small enough so that one could have kept count of the number of objects in question. Josef claims to know that p, where p is that, if there is another object of the same sort to be seen, it will be blue, not red. George denies this. Let us say that Josef is in position Sj, George in Sg.

At first sight it may seem as though Sj and Sg are the same. If, *per impossibile*, that were the case, they would be equally safe, and hence neither would be safe. But, of course, they cannot be identical for George and Josef are making opposing claims: something must account for that difference. Both $\{sj\}$ and $\{sq\}$, the sets of conditions constituting the two positions, include the condition of having observed n objects of a certain sort, each of which was blue. Let $\{d\}$ be the disjoint sum of $\{sj\}$ and $\{sq\}$, thus $\{d\}$ is constituted by the union less the intersection.

If Sj and Sg were equally safe positions, then neither would be safe. But what is important here is that, if they are not equally safe, the difference must arise from and be owing to $\{d\}$. Until we know something about $\{d\}$, it is impossible to say whether one position is safer than the other or whether they are equally safe. And what is wrong with some familiar accounts of induction is that they suggest that of course Josef is in a safer position than George. There's nothing of course about it.

166. It is easy enough to characterize $\{d\}$ in such a way that George is in a safer position than Josef, or in such a way that Josef is in a safer position than George.

Let $\{d\}$ include the condition of having observed n blue and then n red balls inserted in a machine in such a way that the machine would then eject them in the order in which they were inserted. If that were the case, then George might well be in a safer position than Josef. Conversely, if $\{d\}$ were the condition of having observed $2 \times n$ blue balls inserted in the machine, then Josef could be in a safer position than George.

167. In evaluating the relative safety of positions, there are indefinitely many factors that may have to be considered, depending on precisely what p is, on the domain in question and on the character of the attributions in that domain.

If Josef is in a safe position with respect to knowing whether or not he is looking at a rose, it does not follow that he is in a safe position with respect to knowing whether or not he is looking at a member of the family *Rosaceae*: a rose by any other name need not be as well known. Josef may know a rose when he sees one; even so, he need not be knowledgeable about roses: his conception of a rose need not encompass such botanical considerations. But if Josef is so knowledgeable, if he does know that a rose is a member of the family *Rosaceae*, then it is most likely the case that, when looking at a rose,

Josef does know that he is looking at a member of the family *Rosaceae*. There are, however, complications that could appear.

Suppose Josef knows that he has taken an aspirin, and he also knows, having been so informed by reputable authorities, that aspirin is a derivative of salicylic acid. Does Josef then know that he has taken a derivative of salicyclic acid?

One is, I think, inclined to say yes, and perhaps, in a sense, no. For what if Josef has no conception of what it is for something to be a derivative of salicyclic acid, indeed, no conception of what salicyclic acid is? In such a case, Josef might claim 'I know that I have taken, what reputable authorities inform me is rightly classed "a derivative of salicyclic acid", though I have no conception of what such a categorization implies'.

The fact of the matter is that, although we may use the same words, it does not follow that we have the same conception of what is in question. There is no guarantee that, when you say 'I know that there is a giraffe in the garden', and when I say 'I know that there is a giraffe in the garden', what you know is identical with what I know. Indeed, it is highly likely that they are not identical, if what is in question involves areas in which either of us may have developed some expertise. Possibly you might know that there is a giraffe in the garden and yet not know whether there is an animal having an omasum in the garden.

This means that if one is to assess the safety of the position *Ki* is in, when *Ki* claims to know that *p*, one must attend to the character and content of the conceptions being invoked in the claim that *p* is the case. The more detailed and complex the conceptions invoked in the claim that *p*, the more difficult it may be for *Ki* to achieve a safe position with respect to *p*.

If Josef has a simple conception of sheep, and, unlike George, is not concerned with the differences between sheep and aoudads, espying an animal in an enclosure off in the distance, Josef might cheerfully claim to know that there is a sheep in the enclosure; whereas George, in the same situation, would be reluctant to make what might appear to be the same claim. And rightly so: the best that George could claim to know is that there is either a sheep or an aoudad in the enclosure. Suppose that the animal is, in fact, an aoudad, not a sheep: does that establish that Josef does not know what he claimed to know? Let p be that there is a sheep in the enclosure. Then, though it may seem curious, I think one should say that Josef may be in a safe position with respect to p, even though George is not. To say otherwise would be to close ones eyes to the obvious fact that when Josef says 'I know that p', what he is saying is not what George would be saying if George were to say 'I know that p'. If one were to say to Josef 'Look, that's not a sheep, it's an aoudad', his response might well be 'Whatever'. One could, of course, castigate Josef for not being on familiar terms with *ammotragus lervia*: that would not show that he does not know what he claims to know.

168. Suppose both George and Josef are in the position of having noted that, in the past, freshly fallen snow has always been white. Let p be that the next snowfall will also be white. George, in position Sg, claims to know that p, but Josef, in position Sj claims to know that not p. Let the scene be set in a small New England town; let all the conditions be familiar ones, in so far as they can be. Something curious, of course, will have to be the case to account for Josef's claim. But, in so far as possible, everything is to be quite ordinary and familiar. Then of course George is in a safer position than Josef.
Josef, however, argues as follows: 'It is now time t. All snowfalls prior to t were white. To be white prior to t and not white after t is what I call being "grite".[63] All snowfalls have so far been grite: the next one will too'. There is nothing logically wrong with Josef's conception of griteness. (It is not unlike the christian conception of being a being excluded from heaven: to be such, one must either be born sufficiently prior to B.C., or commit certain other crimes.) Furthermore, that Josef had formed such a conception would serve to explain why he claims what he does. Even so, it is perfectly clear that Sg is a safer position than Sj. Why it is can be explained as follows.

Josef has the conception of something's being grite, where to be grite is to be white prior to time t and not white after. I shall say that Josef has the conception of something's being "grite-t" to distinguish Josef's present conception from certain others. Let t-1, t-2, ..., be moments of time prior to t. Let 'grite-t-1', 'grite-t-2', ..., express conceptions identical with that expressed by 'grite-t', save with respect to the specification of the climactic moment. Finally, recall that Josef claims that not p at time t. Since, by hypothesis, all snow-falls prior to t have been white, no snowfalls prior to t have been either grite-t-1, or grite-t-2, and so on. Consequently, the inutility of these conceptions may be taken to be established by hypothesis.

If we now suppose that, even so, grite-t is a useful conception, then we must explain the difference between grite-t and grite-t-1, grite-t-2, and so on, in virtue of which grite-t has achieved some utility. This could be done. That is to say that we could dream up things: t is a special moment in the history of snow, owing to such-and-such factors. But we should have to say what those factors are, we should have to fill out the details, complete, or begin the task of completing, the story in some way. Conversely, if we suppose that grite-t is no more useful a con-ception than any of its predecessors, we have nothing new to explain. Consequently, considerations of coherence direct us to consign grite-t to limbo along with its fellows, andso conclude that Sg is a safer position than Sj with respect to p, that George is in a safer position than Josef.

169. (Since, for reasons which I have never been able to accept, philosophers are often sublimely indifferent to the utility of con-ceptions, it may help to take a short look at this matter from a long perspective.

Whether a particular conception has any utility depends both on the organisms (organic automata) that form the conception and on the ecosystem in which the conception is formed, and in which it is given expression. Generally, and somewhat roughly, speaking, a particular conception is a response to the demands of an ecosystem; the viability of the conception is a matter of how successful a response it happens to be; thus its viability is a reflection of its utility. The formation of a conception is a sophisticated response that few types of organisms are capable of, perhaps only humans and advanced primates at the level of the chimpanzee or above. But the ecosystem, to which it is a response, is, when humans are in question, incredibly complex. The relevant ecosystem for a human being is constituted, not only by his physical environment, but by societal and cultural features of that environment, by the languages, institutions and social practices of the inhabitants of that environment. To appreciate the complex structure of an ecosystem, one should reflect on the fact that it can be so structured that there can be advantages to being dis-advantaged: an apparent inutility may prove to be a concealed utility. In consequence, if one is dealing with a sophisticated conception, such as that of knowledge, it is no easy matter to evaluate the utility of the conception. But that in no way belies the fact that, on occasion, it is relatively easy to perceive changes, which, were they to come about, would serve to undermine the utility of the conception. And when one is dealing with as naive a conception as that of being grite-t, it is evident that it is of no utility whatever.)

170. If Ki, in Si, knows that p, then Si must be safer than any conflicting alternative position, Sj. But merely being safer than a particular occupied conflicting alternative position cannot suffice to establish the safety of a given position.

Suppose Julien, having been asked by a waitress whether he wanted a gin or a vodka martini, was, as he uttered the words, 'I want ...', suddenly crushed to death by a rampaging elephant; further suppose that Julien, a capricious, unreliable and volatile type, enjoyed both types of drinks and frequently ordered either. Then there is no way of knowing, and there never will be any way of knowing, whether Julien was about to order a gin martini or a vodka martini. If Josef, hearing of this, claims to know that Julien was about to order a vodka martini, Josef is in an unsafe position. Perhaps Josef knows more

about Julien than anyone: even so, Josef cannot possibly know what
Julien was about to order. There neither is, nor was, nor ever will be,
a safe position with respect to whether Julien was going to order a
vodka martini. Even so, Josef may be in a safer position than anyone
else to make such a claim. But this must mean that even if one is in
a safer position than anyone else is, or ever will be, it still doesn't
follow that one is in a safe position.

Josef is in position Sj, but it is easy enough to characterize an
unoccupied, but occupiable, position So, such that the union of $\{O\}$,
$\{so\}$ and $\{-p\}$ is as safe as the union of $\{O\}$, $\{sj\}$ and $\{p\}$. For all
one needs is an hypothetical person, as foolish as Josef, who, knowing
what Josef knows, namely that Julien was a capricious volatile type
equally fond of both vodka and gin martinis, is prepared to claim that
he knows that Julien was about to order a gin, not a vodka, martini.

171. What one is confronted with here is the unknowable, the
existence of which may be deplorable, but is not to be denied.

Let p be that there will be people alive on earth in sixty years. The
safest occupiable position with respect to p would be one in which
a person knew, as well as is conceivable, the attitudes of the various
world governments, the potentialities of nuclear weapons, the number
and character of the weapons available, and so forth. If Ki were in
such a position, would he then know whether or not p? Of course not.
We know, as well as we know anything, that it is not possible to
predict the course of human affairs in this way. Some fanatic may
blow the world up tomorrow: there's no way of knowing.

Is it conceivable that someone now know that p, that there will be
someone alive on earth in sixty years? Not at present. However, to
say that it is not presently conceivable is not to say that it is
self-contradictory to suppose that someone knows that p. It is not
self-contradictory, but, even so, it is not presently conceivable. To
conceive of it would be to explain how it could be so, to work out
some of the details. But such a conception cannot presently be
articulated.

What is true is, not that we can now conceive of it, but that we can now conceive of conceiving of it. It is not that in theory we can now say what it would be like to know that p: that we cannot do. No such theory is available. But it is true that in theory we can in theory say what it would be like to know that p, the theory in question being a theory about the powers of human conception.

172. To demonstrate the existence of the unknowable, one need not turn to such momentous matters as whether there will continue to be human life on the planet. There are all sorts of trivial matters that are equally unknowable.

Yesterday, when I fed my cats, I put out a handful or so of cat chow; the bits were countable, but I did not count them. Let p be that I put out exactly twenty three bits. It is quite possible that p is true, but there's no way of knowing. The bits are long since gone, having been ingested and digested by the cats. To make matters worse, when I handed out the bits, I simply added them to a pile already present on the counter, and took them from an already opened box.

173. More interesting examples of the genuinely unknowable are to be located in the domain of the unthinkable.

Human beings are, at best, finite organic automata: there is a finite upper bound to what such automata can process. Just as there is a finite upper bound to the number of images that can be displayed on a television screen,[64] so there is a finite upper bound to the bits of data that the human brain can process. The googolplex is, as far as I know, the largest number that anyone has ever bothered to name.[65] But there are limits to what can be named, and to what can be computed: sooner or later, one runs out of world and time enough. One needn't turn to "inaccessible cardinals" to find the unfindable: the googolplexth prime, in the series of primes, is, quite possibly forever out of reach of computation. That 2, 3, 5, 7, 11 and 13 are all the primes less than 15 is a pleasant thought that one can entertain; but one is forever at a loss when it comes to entertaining the like thought with respect to being less than a googolplex: not even champagne and caviar will turn the trick.[66]

174. One should not, however, confuse the genuinely unknowable with what is unknowable only in that there is nothing to be known. There's no way of knowing the exact number of people in the United States, but that's because there is no such thing as "the exact number of people in the United States". People are being born and dying every few seconds as they straddle the borders. The "number of people in the United States" can only be some statistical result, on a par with the 2 and 1/2 children of the average family.

Just as some matters are not genuinely unknowable, though they may appear to be so, so some matters are not genuinely inexplicable, though they may appear to be so. But the seemingly inexplicable is apt to be more unnerving than the seemingly unknowable. For example, on the day that I have been thinking that, perhaps, I have been acting like a beast, I go to pick up new license plates for my car and find that they are '666 PZ': I have been assigned the number of the Beast in the Book of the Apocalypse. One can, of course, explain why I received the number I did by considering the order in which plates are handed out, the time at which I appeared and so forth. And one can explain why I had been thinking what I had been thinking. But how is one to explain the peculiar character of the coincidence? For the coincidence constitutes a sudden, and here unwanted, apparent increase in local coherence. The correct explanation is, however, that randomness can well have remarkably unrandomlike manifestations.

Perhaps the most beautiful example of this is the "Needle Problem", an experiment performed by the eighteenth century naturalist, Count Buffon:

A plain surface is ruled by parallel lines ..., the distance between the lines being equal to H. Taking a needle whose length L, is less than H, Buffon dropped it, permitting it to fall each time on the ruled surface. He considered the toss favorable when the needle fell across a line, unfavorable when it rested between two lines. His amazing discovery was that the *ratio* of successes to failures was an expression in which *pi* appears. Indeed, if *L* is equal to *H*, the probability of a success is 2/*pi*. The larger the number of trials, the more closely did the result approximate the value of *pi*, even to three decimal places.[67]

175. There may, however, be at least an aspect of the genuinely inexplicable in a merely seemingly inexplicable event. For though one may say that it was a purely random coincidence that I was assigned the number of the Beast on the day that I had been thinking what I had been thinking, one can be inclined to ask: why did that remarkarkably unrandomlike manifestation of randomness occur just then, at that moment of time? And to this there is no answer. There is no way of knowing why it occurred.

Perhaps the right answer to this question is simply that there is no way of knowing because there is nothing to know. If one draws names at random from a hat, it is possible, though unlikely, that they will be drawn in alphabetic order. And if they are, what is one to say? Only that an unlikely event has occurred? Perhaps, but there can be more to it than that.

Here one should consider the well-known so-called "gambler's fallacy", which is, but need not be, a genuine fallacy. Suppose someone is tossing a coin, and one is betting whether it comes down heads or tails. Let *seq* be the sequence of results that have so far occurred. Then, regardless of what *seq* is, the probability that the next toss will be heads remains 1/2: to suppose otherwise is to commit the gambler's fallacy. But suppose *seq* is constituted of a sequence of 17 heads and no tails: should one say 'An unlikely event has occurred' or should one, more plausibly, entertain the conjecture that the tosses are rigged in some manner, perhaps what is being tossed is a coin with two heads? And what if *seq* is constituted by a sequence of a thousand heads and no tails? Sooner or later, one may be forced to call into question the supposedly random character of the event. The gambler's fallacy is, indeed, a fallacy when one is indeed dealing with genuinely random independent events. But whenever randomness has a remarkably unrandomlike manifestation, one at once has some, perhaps ever so slight, reason to wonder whether what appears to be a random event is, in fact, merely that. And the wonderment can provoke a question, 'Why?', to which there may be no answer.

176. Although the thought is, perhaps, repugnant, one must face up to the fact that it is quite conceivable that someone know that something is the case, and yet how he knows it be, at least at present, inexplicable.

George claims to know that Josef is in the theater at Spoleto even though George is in Rome when he makes the claim. When we ask him what makes him think he knows this, he replies 'I gaze into my crystal ball and if I see, or seem to see, a familiar figure in that theater, then I know'. In response to our skeptical smile, George replies 'Call the theater and see if I'm right'. And he is: and what if he is always right? Then it would be hard to deny that he knows what he claims to know, but we wouldn't know how he knows it.

Such a possibility has considerable epistemological significance. For it indicates that the safety of a position need not be indicated by the specific characterization of that position. The primary reason why rational beings reject the claims of crystal ball gazers, psychics, adherents of ESP and others of their ilk, is that the seeming success of such claims has never been sufficiently impressive. More precisely, the meagre reports of marginal success are perfectly amenable to probabilistic explanation: a plausible form of exorcism.

177. There is only one way in which one can positively establish the safety of a given position: that is on the basis of a transference.

Let Sg be the position that George is in when he gazes into his crystal ball; let Si be the position that I am in when I telephone to Spoleto to verify what George has claimed. When I conclude that George is right, I do so on the basis of the assumption that Si is a safe position with respect to what is at issue. Thus the safety of George's position, Sg, is established on the basis of the safety of my position, Si. If I am convinced that Si is a safe position, I may have no option but to concede that Sg is also a safe position. Of course, once I found out that I was forced into such a concession, I would at once be reluctant either to grant the safety of *Si* or to concede that George was, in fact, in position *Sg*. But that George was, in fact, in position *Sg* might then be established on the basis of a transference from some other position, say, the position I was in after a thorough examination of George's premises in Rome, his crystal ball and so forth. And then the safety of *Si* might also be established on the basis of a transference from the safety of some other position, *Sj*, say, the position I am in when I call George in Rome from the theater at Spoleto.

CHAPTER XIV

DEMONS, ANGELS AND MIRACLES

178. EPISTEMOLOGISTS have for years been plagued by Demons, which must be odd since there aren't any. Even so, exorcism is evidently called for.

Imagine, if you can, the following demonic competition: one demon assures me that the world was created one second ago, intact with fossils, pseudo-memories and the like. A second angel (for demons are fallen angels, so we are told,) assures me that the world was created, not one second, but one year ago. In either case, one postulates a miracle, and hence, a naturally inexplicable event. Which is not to say that if an event is, in some sense, inexplicable, it is a miracle; but only that miracles are, in essence, naturally inexplicable. They (would) admit only of supernatural explanation (were there any such).

The supposition that, if I am deceived by an angel, that angel is a demon, is attributable to a deplorable failure to appreciate the place of knowledge in a rational scheme of values. The angel, who deceives me by leading me to believe that the world was created one second ago, does me more good than harm. I should be happy to believe that the misery of the past is merely illusory. The slate then is clean, and the future stands before me, waiting to be despoiled, but still unspoiled. Of course, history may repeat itself. Still, at that moment, the air is clear, the prospects have a seeming brightness: "And love's best glasses reach No fields but are his own". Optimism could abound. The demonic aspect of the would-be deceiving angel is revealed only by her ineffectiveness: she fails to deceive.

The angel who tells me the world was created one second ago is feeble indeed: her wings have been clipped and his horns are showing (for angels have no determinate sex). So are the wings and horns of the

167

second who proposes, not one second, but one year. But there is a slight, ever so slight, alleviation of the miasama that enclouds the first, that serves to enhance the second. For if one is to choose between these alternative views about the duration of the existence of the world, the choice is evident. One must prefer the view, the theory, if it can be so dignified, that creation occurred one year, not one second ago: the number of gratuitous miracles is thereby reduced. For if one second were the right view, how is it that my memories of yesterday coincide with those of some others? My friend recalls that, yesterday, I put the trash out to be collected and it was, and I recall that I did and it was. If the world were created one second ago, this concurrence of recollections of collections would be another miracle. But if the world were created one year ago, the agreement about what was recollected and collected yesterday would be explicable without the invocation of the miraculous.

Rational beings are committed, by their rationality, to the avoidance of gratuitous miracles: to postulate no more miracles than one must is a plausible principle. It is the principle of Rational Reluctance, an heuristic, but not a regulative, principle: it could incline, but not compel, one, for example, to eschew the Big Bang and to reject the Banach-Tarski theorem. Such reluctance is, however, overcome by, but only by, considerations of theory, of coherence: *miracula sine doctrina nihil valent.*

A third angel, who proposes that the world was created, not one second, and not one year ago, but one century ago, easily usurps the ruling position of the second. For again, there is at once an attractive reduction in the number of miracles to be postulated. And so, in this way, one slips and slides into the conclusion that, not seconds, but billions of years, is the best figure as yet figured out by those demonic types who occupy themselves with these questions, cosmologists and astrophysicists. All a philosopher can tell you is to listen to those who are in a better position to know what the truth is about these matters. (But anyone ought to know that such figures as one second, or one year, or even one century, are foolish.)

179. Third rate demons who propose alternative *ad hoc* theories pose no serious problems for epistemologists. But second rate demons are apt to be more unnerving.

A second rate demon breathes the following in my ear as I am strolling peacefully along the beach: 'All one knows', the demon says, 'comes in through apertures in one's surfaces, eyes, ears, nose and mouth, impinges on the periphery of one's nerve endings. (Certainly the world gets on your nerves.) But instead of strolling along the beach, you are, in fact, enclosed in a black box, a box contrived to make you think that you are strolling along the beach, even though, in reality, you are in the box'.

But this second rate demon is a second rate liar: there is, today, no reason to give him credence. Second rate demons merely tittilate one's imagination. No such black box as yet exists. That it could ever exist is far from clear; but that it does not is as clear as can be today. For if it exists, how does it work? The neurophysiological structures of humans are, as yet, largely unknown. One glimpses bits and pieces, here and there. Left and right hemispheres of the brain apparently have marked differences, but the functional plasticity of neural structures is sufficiently impressive to block, or at least impede, easy conclusions. Furthermore, who has produced this black box? How is it operated? What is its source of energy? How does it happen that I am in the box while someone else is outside operating it? Considerations of coherence, of completeness, compel one to turn a deaf ear to second rate demons.

180. A first rate demon can give one pause, for his is a more plausible conundrum. This demon, whose name is "Kreb", has told me this: 'You are, in fact, in a black box, even though you refuse to believe it. You think this is the year 1983. It is not. It is what you would call the year 2777. You were, in fact, born a short time ago; but you were selected to be the subject of the black box experiment. We wished to see how successful we could be in recreating the ecosystem of 1983.

Your refusal to believe me is clear proof that we have, indeed, been remarkably successful'.

The first reply is easy: there is no Kreb, and hence no reason to be concerned. I have simply imagined all this. But then this thought occurs to me, or I imagine this: I think Kreb might say to me: 'You think you are just imagining all this, thinking of conundrums for epistemologists. In fact, in truth, it is I, Kreb, who leads you to think these thoughts by controlling the neural structures of your brain. I repeat, it is the year 2777, a prime time for brain experiments: it is interesting to see how you react to information that appears to conflict with everything you think you know'.

181. Kreb is a first rate demon, I would not deny it. But demons are fallen angels, which means that they are subject to error: Kreb is no exception.

For even if Kreb's claims are correct, and not mere figments of my imagination, they do not, as he suggests, conflict with everything I know. For example, I still know that $2 + 2 = 4$. I also know what I am thinking about, how I feel and so forth. But if Kreb is right, then I don't know many things that I certainly think I know. For example, I know that this is the year 1983. Kreb says that it is the year 2777. If I am right, then Kreb is a mere figment of my imagination. Hence, since I know that this is the year 1983, I also know that Kreb is nothing but a figment of my imagination. On the other hand, if he is right, he does exist, and I don't know what I claim to know. Do I know that this is the year 1983 or don't I?

However, I have told all this to Josef; at first, Josef just laughed, but then, growing more solemn, he said 'I see, Kreb is telling all this to me, not to you: you are just a figment of my imagination'. And he added: 'Kreb is right; I now know that this is the year 2777, and not 1983'.

Let Sz be the position one is in (I am in) in which one claims (I claim) that t, where t is that not f, and f is that there is a Kreb, or he is speaking to Josef, or what Kreb tells Josef is true. Let Sk be the

position one is in (Josef is in) in which one claims (Josef claims) that
f. Then the original question is transformed to two questions: is Sk
as safe, or safer, than Sz, and is Sz a safe position? For if I know that
t, then Sk is less safe than Sz, and Sz is a safe position.

To evalutate the relative safety of positions Sz and Sk both must be
specified, to some extent. What are the relevant differences between
Sz and Sk?

182. Position Sz is the position I am in at the moment, in which I
claim to know that t, that there is no Kreb and so forth. Hence, some
of the truths about me at the moment will serve to characterize that
position.

If what is in question is whether or not there is a Kreb, not every truth
about me that one can think of can be relevant in determining my
epistemic position. For example, that I have brown eyes, rather than
blue, surely doesn't matter. Neither is it significant that I weigh 156
lbs. But this is not to say that all considerations pertaining to my
physical state are irrelevant: if I were subject to spontaneous brain
lesions, then quite possibly I would not be in a position to know what
year it is, depending, of course, on the character and severity of the
lesions. But the specification of position Sk poses more of a problem.
Josef is in position Sk; but if I characterize that position from my
point of view, by saying he is one who gives credence to a fable, I
clearly bias any decision one is likely to arrive at with respect to the
relative safety of the positions involved. (The problem here is somewhat analogous to the problem one encounters in discussions of the
legitimacy of abortion. If one characterizes that which is to be aborted
as a "human being", rather than as a "zygote" or as a "fetus", one
biases the discussion in favor of the illegitimacy of abortion.)

To avoid any bias here, let me simply say that Josef evidently gives
credence to a theory that I reject, namely that he is in communication
with someone called "Kreb", that he has been told that this is the year
2777, and so forth. If so, from the standpoint of Sk, it is perfectly

sensible to credit the view that this is the year 2777, which, however, is not to concede that that is a sensible standpoint.

But there is an odd feature about position Sk that should not go unnoticed: it is not a position that I could be in, even if I wanted to; for to be in that position, conditions $\{sk\}$ would have to be satisfied, but $\{sk\}$ includes the condition of thinking that Ziff is a figment of Josef's imagination. There's no way that I could satisfy that condition: there's no way that I could think that I am a figment of Josef's imagination.[68] Fortunately, there is no need to be in a position to evaluate the safety of that position. Thus one, on occasion, says 'If I were you ...', or 'If I were in your shoes ...' and the like, even though it may be impossible for the one speaking ever to be in such a position. There is, however, another oddity about position Sk when viewed from the standpoint of position Sz, the position I am in: not only can I not be in Sk, I cannot concede that it is a safe position. For to concede that would be to concede that Josef knows what he claims to know, and that would mean that I was conceding that I was a figment of his imagination.

183. Here it is necessary to consider how it is possible to judge that one is in error with respect to p.

Suppose Ki is in position Si, and, in Si, claims, or is prepared to claim, that he knows that p. If Ki is to judge that he is actually in error with respect to p, then he must change his position to one in which he no longer claims, or is prepared to claim, that he knows that p. Hence, he must move into some other position, Sj, that differs from Si, at least with respect to his preparedness to claim that p, for he must now be in a position in which he is prepared to claim that not p. For if he is not now prepared to claim that not p, then he has not judged that he is in error with respect to p.[69]

(Not to digress, one must here take care not confuse judging that one is in error with respect to p with judging that one is in radical error in claiming to know that p. These are radically different matters. One can be in error with respect to p, only if that p be the case is a viable conception. If the attempted claim embodies a conceptual confusion, then though error is at once implicated, it is not an error with respect to the truth or falsity of p. Thus if p is a abbreviation for '17 is divisible by 2 minus the first prime', and Josef says, 'I know that p', then, if, in so saying, Josef is attempting to make a claim, Josef is in error. But the error here is not an error with respect to the truth or falsity of p: p here serves to express a conceptual confusion.)[70]

Assuming that the claim that p does not constitute a conceptual confusion, if Ki is to admit even the possibility of being in error with respect to p, he must admit of the possibility of there being a position Sj that is safe with respect to p and in which one could claim that not p.

But if what is in question is the truth of t, and hence, the falsity of f, I cannot admit even the possibility that there be a position, Sj, that is safe with respect to f, for if f were true then I would be only a figment of Josef's imagination: I cannot concede that Sk might be a safe position.

In contrast, Josef, in Sk, could without difficulty concede that Sz is a safe position, for all that that would imply is that he is, and that I know that he is, giving credence to fables. He would, of course, be conceding that he is in error, but that is not an enormity that cannot be confronted.

184. What is critical here is to realize that one encounters an impossibility in attempting a full comparison of the safety of the two positions in question.

One can conclude that they are equal in safety, in which case neither is safe. Or one can conclude that they differ in safety. Or one can conclude that they differ in safety, and Sz is safe, and hence Sk is unsafe. But one cannot conclude that they differ in safety, and Sk is safe.

The reason for this should not be hard to see. If Sk is a safe position, then from the standpoint of Sk, it is impossible for there to be anyone in Sz; for $\{sz\}$, the set of conditions that serves to constitute position Sz, includes the conditions of being someone who is other than Kreb, and who is speaking with Josef: hence no such position can be occupied, if Sk is a safe position. Josef, in Sk, cannot admit that I am in position Sz since, from Josef's standpoint, I don't exist.

And from this it follows that one cannot, if one compares Sz and Sk, claim that Sk is a safe position: one could so opt for Sk only if there were no need to choose. Thus I could win the argument with Josef, but he couldn't win the argument with me. In this respect, Josef's position is similar to that of a solipsist: he, too, can't win an argument, for if he did, he'd have no one to argue with.

185. The options that are open to us are then the following: the positions are equal in safety, in which case neither is safe; Sk is safer than Sz, in which case neither is safe; Sz is safer than Sk, but neither is safe; or Sz is safe. The problem is to decide between these alternatives. Let us consider the first one.

Are the positions equal in safety? There is no reason to believe it and good reason not to. Position Sz is clearly safer than Sk. Evidence in support of the claim that this is the year 1983 is easy to come by: there is no evidence in support of the claim that this is the year 2777. Josef's testimony is readily explicable in terms of his proneness to fantasy, to excessive indulgence in science fiction and the like.

186. At this point, I imagine Kreb says to me: 'You are deluding yourself with this ridiculous bit of reasoning. The question whether Sz is safer than Sk is simply senseless. To whom can one put such a question? If you put the question to yourself, then, of course, you will conclude that Sz is safer than Sk, for, from your standpoint, position Sk is simply ludicrous; whereas if the question is put to Josef, since he denies your existence, he can hardly accredit the position that you are in. But none of this bears on whether or not what I have told you is true'.

Let's begin again. Let Sz be the position I am in, in which I claim to know that this is the year 1983. Suppose I try to take Kreb seriously, thus I suppose that what he says is true, that he is not simply a figment of my imagination. If I did that, I would then be in position Sz^*. In position Sz^* I would deny that this is 1983. The question then is: is Sz safer than Sz^*?

187. There's some sense to Kreb's complaint: arguing with Josef was senseless, just as arguing with a solipsist is senseless. But Kreb's latest suggestion is hardly original: it's merely an updated version of the antique skeptical claim that, for all I know, I may well be dreaming. Am I dreaming? Of course not. Is Sz safer then Sz^*? Of course it is. But of course 'Of course not' and 'Of course it is' are not arguments, or even reasons, but only claims.

There is, however, a more interesting question to consider than whether or not Sz is safer than Sz^*; that is: is Sz a safe position? It does not follow from the fact that Sz is safer than Sz^* that Sz is a safe position. (That one man is taller than another does not establish that either is tall.) But, if Sz is a safe position, it does follow that Sz^* is not. To answer this question, we must turn to the general question of when one can safely discount the possibility of error.

CHAPTER XV

RISK AND GRAVITY

188. RISKS may, but need not be, encountered in discounting the possibility of error. There is always a possibility of error when one judges that one knows that p, but that does not mean that there is, in fact, always a genuine risk of error. Whether there is a genuine risk of error, and how great it may be, varies from case to case.
There is no risk of error when I judge that I know that $2 + 2 = 4$. But there may be a considerable risk of error when I judge that I know that $a + b + \ldots + j = n$. Thus the risk may be considerable when what is in question is a column of figures such as:

$$
\begin{array}{r}
12349567.45 \\
32948573921.56 \\
45893567553.10 \\
34857494.47 \\
389273.46 \\
29111134.28 \\
921111111.12 \\
\hline
79839960055.44
\end{array}
$$

Is this correct?
If one were concerned with a column of a million figures then it would be virtually impossible to know whether one had arrived at the correct sum, even if one availed oneself of the resources of a huge computer; for there would always be the question whether one had properly keyed in the data.

189. It is hard to say anything of interest in response to a question such as 'How do you know that $2 + 2 = 4$?' primarily because the answer is so obvious.

176

That I am in a safe position with respect to whether or not $2 + 2 = 4$ is simply owing to the fact that I have learned elementary arithmetic. In the course of studying arithmetic one develops a conception of numbers and of numerical relations. Although the etiology of the formation of such conceptions is a difficult subject to research, that people do form such conceptions and, in consequence, are capable of discerning, and do discern, elementary arithmetic truths is not to be doubted.

(A proper discussion of mathematical truth would require more than I can possibly provide here. But a few brief remarks may help to clarify the epistemic and ontologic stance I have adopted.

Knowing such an elementary truth of arithmetic as $2 + 2 = 4$ is essentially a matter of knowing the numbers 2 and 4. For 4 is that number such that that $4 = 2 + 2$ is true, and such that that $4 = 1 + 1 + 1 + 1$ is true and so forth. (Is it surprising that philosophers often find it necessary to state the obvious? For, of course, '$4 = 2 + 2$' serves to express an identity.) And 2 is that number such that that $2 = 1 + 1$ is true. Should the arithmetic sentence '$2 + 2 = 4$' be employed to make a false statement, that would at once cast doubt on the reference of the expressions '2' and '4', or on the significance of " and ' $=$ '. For if the numeral '4' does refer to the number 4, and '2' to 2 and so on, then the arithmetic statement made in saying '$2 + 2 = 4$' must be true. If someone were to say '$2 + 2 = 5$', the reference of the numeral '5' would at once be called into question, which is not to deny that it might prove to be '2' that was tinkered with and so forth.

To the question, 'What is the reference of the numeral '2'?', the obvious answer is that it refers to the number 2. 'What is the number 2?': it is a member of the series of positive integers, it is the only even prime, and so on. Alternatively, one might say that it is a unique infinitary arithmetic structure that may be represented in different ways. Thus one aspect of that infinitary structure may be perspicuously represented by a couple of ducks, or a pair of shoes, or the players in a singles tennis match and so on. The ducks, for example, perspicuously represent that aspect of the structure indicated in the identity $2 = 1 + 1$, for a couple of ducks is one duck and another duck.[71] Should one of the ducks fly off, and one be left with only one duck, the remaining duck would perspicuously represent a particular aspect of the structure of 1, namely that indicated in the identity $1 = 1 + 1 - 1$. And what if there had been only one duck to begin with? 'Count the ducks!' makes sense in a way that 'Count the duck!' doesn't. Even so, one might be told to count the ducks and then find that there is only one duck. If so, the presence of the single duck would then constitute a perspicuous representation of an aspect of the structure of 1 indicated in the identity $1 = 1 + 0$. Would the presence of a single duck also constitute a representation of that aspect of the structure of 1 indicated in the identity $1 = 97 - 96$? Of course it would, but it would not, save in special circumstances, constitute a perspicuous representation. How can something be a perspicuous representation of an aspect of something, a, and yet not a perspicuous representation of an aspect of something, b, when a is identical with b? Consider two representations of a triangle, one of which is marked to indicate that the interior angles are all equal, the other of which is marked to indicate that the sides are all equal. Then the first is a perspicuous representation of an equiangular triangle, while the second is a perspicuous representation of an equilateral triangle.

Representation is not exemplification: mathematical entities (structures) cannot be exemplified. A line on a blackboard does not exemplify a mathematical line in the way that a green leaf exemplifies greenness. But mathematical entities (structures) can, nonetheless, be represented; thus a chalk line on a blackboard can represent, even though it does not exemplify, a mathematical line. And one can acquire knowledge by means of representations. To speak somewhat loosely, if one thinks of elements of symbolic systems as representations, then all my knowledge of, say, China has been acquired on the basis of representations. The same is true of my knowledge of Ancient Babylonia. One could, perhaps, go to China to confirm

matters. But no one is going to Ancient Babylonia: that doesn't mean
that one can't know anything about Ancient Babylonia; we do, all
sorts of things.

Kronecker had something of a point when he said that God made the
integers, man the rest of mathematics. But what he should have said
is that we become acquainted with the integers through representations, whereas our knowledge of the rest is achieved through
deductions, constructions and the like.)

190. There would be a genuine risk of error were I to judge that I know
that my car is in my garage. I am upstairs in my study, my garage is
downstairs and out of sight from the study; furthermore, I have not
looked into the garage for several minutes: even so, I know that my
car is in the garage. The risk of error, in this case, can be seen in the
fact that it is easy enough to characterize a not implausible hypothetical alternative that could serve to defeat my judgement that I
know that my car is in my garage.

Let p be that my car is in the garage; let Si be the position I am in;
let Kh be an hypothetical person, who is in position Sh, that is
constituted by a set of conditions, $\{sh\}$, that includes the conditions
of having entered my garage, having moved the car out of the garage
onto the driveway and so forth. Then it is conceivable that the union
of $\{O\}$, $\{sh\}$ and $\{-p\}$ prove to be more coherent than the union of
$\{O\}$, $\{si\}$ and $\{p\}$.

191. It is here that one finds the *locus classicus* of the problems of
epistemology: how is one to assess the impact of a purely hypothetical
alternative on an open system? For if the construction of an hypothetical alternative, Sh, is to establish that Si is not a safe position,
the union of $\{O\}$, $\{sh\}$ and $\{-p\}$ must be no less coherent than the
union of $\{O\}$, $\{si\}$ and $\{p\}$.

If one says 'I know that p', a skeptic replies 'But what if ...?'; thus
he counters with the construction of an hypothetical alternative. And

sometimes the hypothetical alternative suffices to establish that one doesn't know what one claims to know, and sometimes it doesn't.

192. To appreciate the problem here, one must appreciate the difference between actual alternatives and hypothetical alternatives.

When one is dealing with actual alternatives, there is no theoretic difficulty in making comparisons of relative safety. The domain of speculation is limited to the actual world: lupine fancies are kept at bay. But unleashed hypotheticals know no limits.

I know that my car is now in the garage downstairs; I parked it there just a few minutes ago. But what if, in those few minutes, thieves mangaged to enter the garage and remove the car? Is that possible? Not really; but it is conceivable. However, I would have heard them. But what if they managed to do it silently? Is that possible? Again, not really, but, again, it is conceivable. However, no one would attempt such a thing in broad daylight. But what if they were adventurous thieves? And so on. Or, to take another tack, what if I am in error about having parked my car in the garage? Possibly I meant to, but, in fact, I left it in the driveway. What if I have just had a lapse of memory? I've been working too hard lately, and such things are known to happen. And so on. Do I know that my car is in the garage? I do indeed. But what about the hypothetical alternatives that have just been sketched? They prove nothing.

There are limits to the hypotheses that one is prepared to entertain in searching for an hypothetical alternative that can serve to defeat the judgement that one knows that p: the limits depend on what one is prepared to discount.

193. Discounting factors is an essential aspect of all abstract judgements, and all judgements have an abstract character.

If I judge the 'a' on the left hand side of the page to be the same as the 'a' on the right hand side of the page, I therewith discount the spatiotemporal and graphological differences between them. If I judge milk to be a healthful food, I discount the fact that milk laced with strychnine is still milk but certainly not healthful. If I judge an El Greco *Crucifixion* to be a good painting, I discount the fact that I, personally, am somewhat put off by what I take to be its saccharine religious aspect.

By discounting various factors, one may achieve a certain synthesis, one may manage to organize and integrate one's behavior, attitudes and beliefs.

194. Discounting possibilities is a way of life: it is the only way, if one is to live and act as a rational being. There is no option here.

There is a possibility, certainly remote, but a possibility nonetheless, that the food I have prepared for my evening meal will poison me; and that possibility will remain no matter how many tests I might make. (I know little of such tests.) Realizing this, reflecting on this, should I eat my meal in a spirit of resignation, or with an air of bravado, or with indifference? Surely none of these would be a rational attitude: instead, given that one has obtained one's produce from familiar and hitherto reliable sources, one simply discounts such far-fetched possibilities.

Driving over a bridge at night, one may realize that there is a possibility that a section of the bridge collapsed during the day and that, in fact, there may be no roadway up ahead, at a point just out of reach of the car's headlights. Or, for that matter, one may realize that there is a possibility that the bridge will collapse as one is driving across it. (Such things have, indeed, happened.)

Or walking by an apartment building in Chicago, one may realize that there is a possibility that the entire building will explode and blow up the neighborhood. (Again, such things have happened.)

Or should I test every chair before seating myself in it, for it is possible that it will collapse under my inconsiderable weight? But then, perhaps I should test the floor, to see if it will hold me while I am engaged in testing whether or not the chair will hold me. And then there is the possibility that my tests are not adequate, and perhaps I should devise tests to test my tests.

The possibility of sudden death and disaster is always with us; but that doesn't prevent a rational being from making plans to eat a meal, play tennis, take a trip to the beach. When I announce to my guests 'Dinner will be ready in a moment', I discount the possibility of my demise before the soup can be served. (Less morbid types are no doubt simply oblivious to that possibility. Perhaps being oblivious is the better way.)

195. Discounting a possibility is not the same as taking the possibility into account and acting accordingly.

Reflecting on the possibility of a fatal crash, an airline passenger may decide to fly nonetheless; but if, in consequence of his reflection, he elects to purchase life insurance for the trip, he has not discounted the possibility of a crash: on the contrary, he has taken it into account and taken a calculated risk.

If one refused ever to discount possibilities then, indeed, "the native hue of resolution" would be "sicklied o'er with the pale cast of thought": one would be paralyzed, immobilized. There is a risk involved in driving to the market to buy groceries; I take this risk whenever I drive: it is not always a calculated risk. Often I would not know how to calculate it, and who has time and energy both to calculate and to shop?

196. One may profess to believe that p or one may claim to know that p; but in either case (assuming that both the profession and the claim are sincere), one is either discounting the possibility of error or merely oblivious to it.

If there is a risk involved, there is less of a risk in professing to believe that p than in claiming to know that p. There may be a risk nonetheless. If someone claims to know that the problem of inflation will disappear overnight, he is in error. But if someone sincerely professes to believe that the problem of inflation will disappear overnight, he surely has a false belief and, to that extent, is also in error. Even more, he has an unreasonable belief. It is not as though retreating to professions of belief, and thus avoiding claims to knowledge, could serve to obviate the necessity of coping with the possibility of error. Unreasonable beliefs and unwarranted claims to knowledge are alike in this respect: in either case, one discounts or is oblivious to a possibility of error that cannot safely be discounted.

197. A profession of belief may be less risky than a claim to knowledge, but the acquisition of kowledge is a greater achievement than the formation of a true belief.

If I truly believe that p, but I do not know that p, my true belief, nonetheless, constitutes an increase in coherence. For, if I believe truly that p, then I have considered whether or not p is the case, and, in the absence of conflicting considerations, I am prepared to state that p; in which case, there is a conformance between the statement that I am prepared to make and the conditions that actually obtain. But if I know that p, and not merely believe truly that p, then further conditions must be satisfied. Let $\{k\}$ be the further conditions that must be satisfied. Then the difference between believing truly that p and knowing that p is that, in the former case, there is a conformance between what I am prepared to state and the actual satisfaction of the truth conditions for p, whereas, in the latter case, there is an accordance, not only between what I would state and the actual satisfaction of the truth conditions for p, but between the requisite conditions, $\{k\}$, and the conditions that actually obtain. In this respect, knowing that p constitutes a greater increase in coherence than merely believing truly that p.

The character of the set of conditions, $\{k\}$, the conditions that must be fulfilled if one is to have knowledge, and not simply a true belief, deserves serious consideration, for it has a truly striking character.

198. There is a correlation between the difficulty of fulfilling $\{k\}$ and the resultant increase in coherence constituted by the achievement of the knowledge in question.

(Again, there is a remarkable analogy between knowledge and negentropy: an increase in the difficulty of fulfilling a set of conditions is

analogous to an increase in the amount of energy to be expended.)
If I wish to know whether or not the sun is shining at the moment,
all I need do is glance out the window. But if I wish to know whether
or not 30103 is a prime number, a great deal more effort may have
to be expended.

Consider two cases: in the first, I simply consult a table of prime
numbers; in the second, I determine the answer by performing a
computation. Let $\{k1\}$ be the set of conditions fulfilled in the first
case, $\{k2\}$ the set in the second case. Then, clearly, the fulfillment of
$\{k2\}$ poses a greater difficulty than the fulfillment of $\{k1\}$. Since what
is in question is a difference between different ways of achieving
knowledge, by hypothesis, in either case p is true and one achieves
the knowledge in question. But if one achieves the knowledge on the
basis of a correct computation, thus by fulfilling conditions $\{k2\}$, one
has achieved a greater coherence. This should be obvious if one
reflects on the systematic character of the act of computing whether
or not 30103 is a prime, in contrast with the casual character of the
act of consulting a table of primes.

199. Whether a possibility of error may safely be discounted depends
on the factors of risk and gravity.

There is no metric available to express the measure of risk involved
in a claim to knowledge; but, intuitively speaking, it is reasonably
clear that some claims are riskier than others. Given that I am not
an aphasic, less risk is involved in claiming to know that my car is
in my garage when I am in the garage looking at a car than when I
make the same claim upstairs in my study, the garage being out of
sight.[72] And if I were to claim to know that the courts at the local
tennis club are now playable, despite the fact that it rained last night,
that would be an even riskier claim than either of the others.

Conversely, the element of risk is minimal, or even nonexistent, when
I profess to know that spiders have eight legs, that the Moon orbits
the Earth, that I will die and so forth.

But it is not, I think, generally possible to make detailed precise comparisons of risk. Is there a greater risk involved in claiming to know that hippopotami feed along a river bank but defecate in the river than in claiming to know that giraffes tend to browse on acacia leaves? Both facts have been established on the basis of prolonged observations of the behavior of the animals in question. The observations in the case of the hippopotami were more difficult, for they necessitated underwater operations; but an increase in difficulty does not imply an increase in risk of error. The observations in the case of the giraffes had to be sufficiently prolonged and varied to establish that the observed behavior was neither a momentary aberration nor an eccentricity of the particular herd being studied.

200. Whether there is a genuine element of risk in a particular claim to knowledge may be somewhat of a subjective judgement.
I recall sitting in on a seminar in which the philosopher performing denied that "I knew for certain" that there was an ashtray before me, on the seminar table. There was an ashtray in front of me, I had picked it up and banged it on the table. I had claimed to know that there was an ashtray before me. (The locution 'know for certain' is not part of my idiolect, but insofar as I understand what was being said, I was claiming that I "knew for certain" what was in question.) As far as he was concerned, the risk of error was so considerable as to preclude the possibility of my rationally claiming to know that there was an ashtray before me. As far as I was concerned, there was no risk of error at all, and hence nothing to deter me from claiming to know what I did in fact know, namely, that there was an ashtray before me. Was there any risk of error?
If there is to be a risk of error, the risk of error must have some source. What, in such a case, could it be? There was no risk of a conceptual confusion: I know what an ashtray is, what a table is, what it is for an ashtray to be on a table and so forth. Was there a risk of my having or being subject to an hallucination? I find that thought simply ridiculous. I have never had an hallucination in my life. Anyway, if I were to have an hallucination, why should I hallucinate an ashtray?

A pink rat, a glass of vodka, a tennis ball, a warm stretch of sandy beach would be more likely candidates. Furthermore, even though no one else claimed to know that there was an ashtray on the table, all professed to believe that there was. Could it have been, not an hallucination, but an holographic image of an ashtray projected on the table? One cannot bang a holographic image on a table.

201. The gravity of the judgement that one knows that p is determined by a variety of factors; but the central theme is always this: the effect on coherence.

If p is itself trivial, of no consequence to anyone, the judgement that I know that p is then equally trivial. Thus, if I am not suffering from any visual impediment, or any ailment pertaining to the eyes, were I to judge that I know that I blinked my eyes a moment ago, the judgement would be utterly trivial.

Suppose, however, that the fact about the hippopotami had not been known. Then one who judges that he knows that hippopotami feed along the river bank but defecate in the river is making a judgement that could have a considerable effect on his understanding of the ecology of the river. For the hippo's behavior would constitute a transference of nutrients from the bank of the river to the river bottom; the fertilization of the river bottom would give rise to the growth of underwater organisms; these organisms would, in turn, become a food source for insects, the insects for fish, the fish for birds, and these in turn would attract still other predators.

202. There may be a risk of error in judging that one knows that p, but some risks are worth taking, some are not.

There is a risk of error in judging that one knows how hippopotami behave, but if one is not in error, if one does in fact know what is in question, the appreciation of that knowledge could greatly effect one's understanding of the ecology of a river. And that means that such knowledge could be conducive to a considerable increase in overall coherence.

On the other hand, given the gravity of the issue, should the judgement prove to be in error, the error could be conducive to a considerable decrease in overall coherence.

203. The gravity of a judgement that one knows that p should not be confused with the gravity of a claim to know that p: these are different matters.

To claim to know that p is to perform a social act; it is to engage in a specific social practice in which one person provides assurance to another. If you ask me whether I know that p, then why you want to know whether or not I know that p can, in some cases, determine whether or not I can reasonably claim to know that p. This is not as weird as it may sound; for whether or not it is reasonable to commit oneself to something may depend on precisely what one is committing oneself to. Even so, from the standpoint of traditional epistemology, the view that I am suggesting must seem pretty far out. The traditional view would certainly be that either I know that p or I do not; whether anyone wants to know whether or not I know that p is another, and irrelevant, matter. And with that I certainly agree. But the traditional view also seems to be that if I know that p then I should be prepared to claim that p, and if I do not, then I should not, and with that I do not agree. What I am saying is that it is one thing to know that p: it is another to be prepared to claim to know that p. Even if I know that p, it may not be reasonable for me to claim to know that p in one case, and yet reasonable to do so in another.

If p is not itself trivial, if p is a well-known fact, say, that the Earth is part of the solar system, the claim to know that p may still be rather trivial. How trivial it is depends on the situation in which the claim is made and to whom I am speaking when I make the claim. For me to make the claim in speaking to a competent astrophysicist would simply be ludicrous. There could, however, be some point in making such a claim if one were speaking to a child, or to a remarkably uninformed person.

If p is a hitherto unknown fact, the gravity of the claim to know that

p may be considerable. How grave the claim is will depend on its ramifications, or, more precisely, on its effect on overall coherence. A crucial aspect is the social transmission of the claim: if the claimant has no appreciable audience, if his audience gives him no credence, then the claim is not likely to have any significant effect on overall coherence. Thus any claim I might make about astrophysical matters is likely to be utterly trivial: who would listen to me? On the other hand, if a doctor claims to know that I have a terminal cancer, whether he is justified in making such a claim depends, not only on whether he knows what he claims to know, but on the consequences of his making such a claim. His claim is, so to speak, in double jeopardy; if he is in error, his claim to knowledge will have compounded the error.

204. (The distinction between judging that one knows that p and claiming to know that p is, I am inclined to think, altogether obvious. Why then bother to draw attention to it?

Why has anyone ever raised the question whether knowledge might possibly be justifiable true belief? Justification is a social conception, and so, of course, is the conception of justice. Justification may be required if one claims to know that p, but it has nothing whatever to do with judging that one knows that p.

Knowing or believing are not things one can choose. If Josef says 'I choose to believe George rather than the police', what he is saying is that he gives greater credence to what George says than what the police say; but his gift of credence is not a matter of choice. And if Josef says 'I refuse to believe the police', what he is saying is that he will not attend to what the police say in any way that is likely to result in his giving credence to what they say. One can choose to attend or not to attend, at least on occasion, but one cannot choose what the upshot of that attention may be. And since believing or not believing that p is not something that one can elect to do, it is simply senseless to speak of "a justified belief", unless that is construed either as an ellipsis for 'a justified profession of belief' or as a peculiar variant of

'a reasonable belief', or a 'well-founded belief' or something to that effect.
Neither knowledge nor belief either require or can even accept justification.)

205. There is a complex relation between considerations of risk and considerations of gravity in judging that one knows something.
If the gravity of a claim is minimal, we are prepared to tolerate an increased risk: the requirements for knowing that p are, accordingly, relaxed. In this respect, the conditions requisite for knowing that p are somewhat analogous to the defeasibility conditions for promises. Suppose I promise to meet you for a drink at the local bar, but, having severely sprained my ankle on the way, I decide to return home, and hence fail to show. My excuse, that I severely sprained my ankle, is likely to exonerate me. But suppose that you are in the hospital, desperately in need of a blood transfusion, that you have a rare blood type and I am the only available local donor. I promise to come to the hospital at an appointed hour, but, on the way, I severely sprain my ankle. If I were then to decide to return home, and hence fail to show, my excuse would not serve to exonerate me.

206. The reasons why there is such variability, both in the defeasibility conditions for promising and the conditions requisite for knowing that p, are reasonably transparent.
In each case, it is a matter of utilities. A cost-benefit analysis of the practice of making promises would indicate that a tightening of the defeasibility conditions could correspond to a decrease in the utility of the practice. One simply would not make a promise to meet someone for a drink, if spraining one's ankle on route would not constitute a legitimate excuse for failing to perform.
There is, of course, a problem of balance here. An excessive relaxation of the defeasibility conditions for promising could also correspond to a decrease in the utility of the practice. If virtually anything would get me off the hook, then my promise would be worth virtually nothing. Analogously, if virtually any position would allow the judgement that

one knows, the utility of the conception of knowledge would be virtually nothing.

207. Consider two situations, in each of which I put to myself the question whether I, in fact, know that my car is in the garage, and in each of which I judge that I know that it is.

In the first case, I am aware of an irritable son in want of a tennis racquet which he wishes to repair and which is in the car. If the car is in the garage, he can gain access to it through the house; whereas if it is in the driveway, he would have to cope with mud and rain, in which case it would not be worth the effort. He mistakenly believes that the car is not in the garage and complains about my having carelessly left it out.

In such a case I would cheerfully claim to know that the car is in the garage even if my only basis for so claiming is a recollection of having parked it there last night. The element or risk may be appreciable, but given the triviality of the claim, I cheerfully risk it. Should I prove to be in error, the error would result in merely a momentary perturbation (and perhaps a temporary reduction in my credibility rating with my son).

In contrast, if what is in question is the availability of the car to transport someone to a hospital at a moment's notice, before judging that I know that it is in the garage, I would go down to take a look and see that all is secure and ready. For, in this case, should I be in error, the error would be conducive to considerable incoherence, indeed, local chaos.

208. These two cases require fairly careful attention, for they have a somewhat misleading aspect. Since in each case I judge that I know that my car is in the garage, it may seem as though the gravity of the judgement must, in each case, be precisely the same: but that is not the case.

I said previously that, by discounting factors, one may achieve a certain synthesis, one may manage to organize and integrate one's behavior, attitudes and beliefs. This means that, by discounting factors, one may manage to achieve a greater coherence.

If, in the first case, my judgement proves to be correct, the car is, indeed, in the garage, the resultant increase in coherence may be manifested in close in simple ways. My son is likely to proceed to the garage, find the car, remove the racquet and, thus, behave as I expect. If, however, I should prove to be in error, the resultant decrease in coherence is likely to be minimal: the car will not be in the garage, my son is likely to be further irritated, I shall deem myself to have been subject to an aberration, but that's about all.

In the second case, if I am not in error then, again, matters are likely to proceed much as I expect. But if I should prove to be in error then, unlike the first case, my expectations about how events will proceed, how people will behave and so forth may be thoroughly upset: matters may be utterly unpredictable.

209. Unlike the defeasibility conditions for promises, the conditions requisite for knowledge are subject to constant correction.

The defeasibility conditions for promises, obligations and the like are, as is everything else in this world, subject to alteration in time. But there is no clear logic to the process of alteration. Our moral consciousness is susceptible to change; as it changes, so do defeasibility conditions. If you fail to perform a promise to meet with me for a drink, and your excuse is that you had to care for an injured rabbit in your garden, would that excuse serve to exonerate you from all blame? It would with me. It would not with some others, who can remain unmentionable. What if your excuse were that you had to care for a displaced harvestman, a splendid Daddy-Long-Legs? Again, that would be an acceptable excuse to me, but, I am sure, not to many others.

Today one can discern a growing concern for the rights of animals; and so, one might speak of a growing "recognition" of the obligations men have to beasts. But such "recognition" is, in truth, more akin to creation than to realization. Our moral consciousness is constantly undergoing a transformation: new rights, duties and obligations keep appearing, and our moral scheme of values is subject to constant fluctuation. There is then a corresponding fluctuation in the defeasibility conditions for promises.

What is of importance here, however, is the fact that the defeasibility conditions for promising are not subject to correction simply in virtue of the fact that they have been invoked. If I fail to perform a promise, but I have a legitimate excuse for failing to do so, that excuse does not, *ipso facto*, henceforth cease to be a legitimate excuse. Whereas if in position *Si* I judge that I know that *p*, but *p* proves not to be the case, henceforth *Si* must be deemed to be an unsafe position.

It is this fact, perhaps more than any other, that warrants one in appealing to an open system in speaking of knowledge.

210. One's conception of one's knowledge may admit of corrections, but there is a limit to the corrections that it can sustain.

Suppose one morning I awake thinking that, since today is January 23rd, perhaps I should have a dinner party tonight. There was a heavy snowfall yesterday, but the roads should be clear enough to permit travel, guests could arrive. As these thoughts go through my mind, I suddenly realize that there is something odd about the bed I am in; it is not my bed: I am in a hospital bed with bars on each side. At that moment, a person in an orderly's uniform enters the room. A few moments later, friends and relatives appear. Gradually the story unfolds: I am in, and have been in, a mental institution for the last six months.

It is not January but July; during that period I suffered from extreme delusions: I had proclaimed that I was Prince Hamlet, that I was in Denmark, that Rosencrantz and Guilderstern, being two, were three, and so on. Of all this I have no recollection whatever; but on looking through the barred windows I see, not snow, but everything in full bloom: the crepe myrtles are heavy with blossoms, and they do not blossom here before July. Looking into a mirror, my hair displays at

least a six months growth. And so on: everything that I can see and
hear confirms what I have been told.

Do I then conclude that, yes, I have been quite mad for the last six
months? I don't know. Could I conclude anything? Perhaps now I
am mad. Thank Humpty Dumpty this has never happened to me!
If I know that something is so then it is so and I am in a position in
which the possibility of error may safely be discounted.

KREB'S EPILOGUE

AT MY BACK I hear Kreb intoning: "This is the year 2777, though you refuse to believe it. You are in a black box in which all your thoughts are controlled, all your perceptions, all your feelings. The experiment is a complete success".

No, there is no Kreb: he is simply the product of my imagination. I know that this is the year 1983. There is no evidence to support Kreb's view; there is no reason to believe that he exists.

"For, even in 1983, great advances had been made in neurophysiology; even in 1983 the possibility of a complete computerized brain scan had become more and more obvious. And yet so-called "modern" technology was still in its infancy".

I know that there is no Kreb, but what if I were wrong? I am not, but I could be, but I am not, though I may be.

A wall has been built, and it is being built; we think it will continue to be built. No one knows exactly who started the wall, though many helped. Nor does anyone know how far it reaches: it seems to go on and on forever. We think the builders are our principals.

The wall is to protect us from the invasion. Wall soldiers man the wall. Whenever a soldier is overcome by an invader, he must be replaced by a stronger soldier, and we are forever sending replacements. We have even sent soldiers to man the wall in the distant provinces. But no one knows how strong the enemy forces are there. We need as many soldiers as we can get, but we want only those who are strong enough to repel an invader. It is possible that there is a man who is strong enough to repel an invader. We can find out whether one man is stronger than another. We know if a man isn't strong enough if he is overcome by an invader. But if he is not, we don't know whether it is because he is strong enough, or good fortune has kept

194

stronger invaders away. We have found a section of the wall where the invaders are too strong for anyone weaker than K. So we know that no man weaker than K will do there. For the time being, we risk it: we judge that K is strong enough. Perhaps someday K may have to be replaced. Yes, we know that.

Meanwhile we stare at the long reaches of the wall and wonder.

NOTES

[1] For a detailed discussion of meaning and related issues, see my *Semantic Analysis* (Ithaca: Cornell University Press, 1960).

[2] See G. Ryle, *The Concept of Mind* (London: Hutchinson's University Library, 1949) 152.

[3] Again, for details, see my *Semantic Analysis*, sec. 181 ff., but also my *Understanding Understanding* (Ithaca and London: Cornell University Press, 1972) Chapter II.

[4] Of those who cannot hear it, I can only say, to quote T. Patton, "the metal of their ear is showing": a tin ear is, perhaps, a blessing in marital life, but not in philosophy.

[5] See my *Understanding Understanding*, 21–38, for further discussion of vectors in linguistic analysis.

[6] For a discussion of coherence, see Chapter IV, below.

[7] For a discussion of reference, see Chapter III, below.

[8] See my "The Nonsynomymy of Active and Passive Sentences", in *Philosophic Turnings* (Ithaca: Cornell University Press, 1966) 147 – 154.

[9] T. Givón, *On Understanding Grammar* (New York: Academic Press, 1979) 188.

[10] For a detailed account of these and similar matters, see E. Sapir, *Language* (New York: Harcourt Brace and Co., 1949) originally published in 1921. In particular, see 28–29.

[11] See Chapter II of my *Understanding Understanding*, and see below.

[12] *Speech Acts* (Cambridge: Cambridge University Press, 1970) 77.

[13] *Op. cit.* 78.

[14] *Op. cit.* 79.

[15] *Op. cit.* 78.

[16] *Op. cit.* 79.

[17] *Op. cit.* 78.

[18] *Mathematical Structures of Language* (New York: John Wiley & Sons, 1968). 7. Harris' footnotes, here deleted, are not relevant in the context of this discussion.

[19] But though I am a firm admirer of Skinnerian technology, I must confess that I am not one: I do, however, like to parade in such clothing.

[20] But these matters admit of enormous complications: see my *Understanding Understanding*, Chapter II, for example. For an account of how conditions are associated with expressions, see my *Semantic Analysis*. For an account of proper names, see my "About 'God'", in *Philosophic Turnings*, 93–102; also see my "About Proper Names", *Mind* 86 (July, 1977).

[21] I am indebted to Robert Brandon for criticisms, and to R. Michael Harnish for suggestions, that have served to improve this discussion of coherence.

[22] See my *Understanding Understanding*, Chapter III "The Logical Structure of English Sentences" 39–56, for a complementary perspective on the same issues.

[23] *The Problems of Philosophy* (New York: Henry Holt and Co.) 73.

[24] *Ibid.*

[25] Laver is called 'Rocket' by his intimates; Roy Emerson is called 'Emmo' by his intimates.

[26] *Cf.* his *On Understanding Grammar* 278–289.

[27] *Op. cit.* 278–289.

[28] See Chapter VII, below, for an explanation of where Givón has gone wrong.

[29] See Chapter VII Knowing How, below, for further discussion of internalization.

[30] Again, see Chapter VII Knowing How, below.

[31] See Chapter XV Risk and Gravity, below.

[32] And, as an aside, let me say that thinking in Italian is not the way to do philosophy: not that I am an Empiricist, but Italy could never be the home of Empiricism; the genius of the language is abstract, romantic, imprecise and enormously seductive: live Italian, by all means, but do philosophy in English!

[33] See my "The Number of English Sentences", *Foundations of Language* 11 (1974) 519–532.

[34] This was pointed out to me, years ago, by John Austin in conversation: my indebtedness to Austin, to his preoccupation with precision, should be apparent on every page of this essay.

[35] As a child I could not refrain from constantly interjecting 'I don't know' by way of a futile protest against what I took to be an abuse of the word 'know'.

[36] Perhaps this is the source of error in N. Malcolm's claim that "If you lack confidence that p is true then others do not say that you know that p is true, even though *they* know that p is true. Being confident is a necessary condition for knowing". See his *Knowledge and Certainty* (Englewood Cliffs, N.J.: Prentice-Hall, Inc., 1963) 59–60.

[37] I am well aware that this view is contrary to the received opinion among epistemologists, but so much the worse for the received opinion. However, see Chapter IX Conditions, for a further discussion of this matter.

[38] A. J. Ayer, *The Problem of Knowledge* (New York: St. Martin's Press, 1965) 34.

[39] W. V. Quine and J. S. Ullian, *The Web of Belief* (New York: Random House, 1970) 3.

[40] *Op. cit.* 4.

[41] I owe this example to S. Munsat.

[42] See my "A Response to "Stimulus Meaning"" in *Understanding Understanding*, 90–106.

[43] *Cf.* his recent book, *Philosophical Explanations* (Cambridge: Harvard University Press, 1981).

[44] *Op. cit.* 178.

[45] A character on the street: "I've hear of T. S. Eliot; he wrote a poem called *Four Quarters* and talked about "Dime past and Dime future"; he must have wrote during the Big Depression".

[46] Perhaps one should say that, not only can one hold the right view for the wrong reasons, but one can hold the wrong view for the right reasons, which seems to model the view that the road to Hell is paved with good intentions.

[47] John Cook-Wilson, *Statement and Interference* (Oxford: Clarendon Press, 1926) 105–106.

[48] *Op. cit.*

[49] See the many writings of Noam Chomsky.

[50] Perhaps I should stress that my remarks here do not constitute, nor do I conceive of them as, an objection to Chomsky's real views: my objection is to his rhetoric. Chomsky is well aware of all the points I have raised. (I know this because we have discussed these matters.) But he is also evidently aware of the fact that it is vastly more provocative to speak of "innate knowledge" than to speak, more precisely, of innate neurophysiological structures which, under appropriate conditions of maturation and subject to appropriate stimulation, are likely to give rise to knowledge. He chooses to be provocative: provocative rhetoric, not precise statement, starts movements and storms barricades. Slogans, not precise statement, carry the day: *esse ist percipii*, existence precedes essence, to be is to be the value of a variable, etc. Why do I object? Alas, I have no slogan; instead, a possibly foolish commitment to the truth.

[51] *Cf.*, for example, Jaakko Hintikka, *Models for Modalities* (Dordrecht, D. Reidel, 1969).

[52] *Cf.* my *Semantic Analysis*, 44.

[53] See Chapter XII, Skepticism, below.

[54] See Chapter XIII, A Safe Position, below.

[55] And from this standpoint, one can see the truth that aestheticians who spoke of "suspension of belief" evidently caught a glimpse of, but misdescribed.

[56] Quoted in J. R. Newman, *The World of Mathematics* (New York: Simon and Schuster, 1956) 375.

[57] As is the value of art, and indeed, all intrinsic values. See my "Art and Sociobiology", *Mind* (1981) XC 505–520.

[58] *Atomic Physics and Human Knowledge*, (New York: Science Editions, Inc., 1961) 77.

[59] From Colin Cherry, *On Human Communication* (New York: John Wiley & Sons, Inc., 1957) 215.

[60] Harlow isolated monkeys at the bottom of a giant funnel, which he (or someone) baptised "The Well of Loneliness", to see whether the animals would be psychologically devastated: they were. It would not be difficult to find an appropriate human subject for the same experiment.

[61] Such as Tertullian: "Wretched Aristotle! who taught them dialectic, that art of building up and demolishing, so protean in statement, so far-fetched in conjecture, so unyielded in controversy, so productive of disputes; self-stultifying, since it is ever handling questions but never settling anything...". From H. Bettenson, ed.

Documents of the Christian Church (New York & London: Oxford University Press, 1947) 9–10.

[62] This was pointed out to me years ago by M. White, who, however, added that humor has no place in philosophy.

[63] The word is suggested by Goodman's "grue". See N. Goodman, *Fact, Fiction and Forecast* (Indianapolis-New York: Bobbs-Merrill Co., Inc., 1973). The problem to be described is essentially Goodman's problem of projectible predicates, and the resolution of it in terms of coherence is not incompatible with Goodman's own resolution in terms of "entrenchment"; however, entrenchment is merely a manifestation of coherence.

[64] Some years ago the number was in the neighborhood of 2 to the 186,000th power; possibly it has since increased.

[65] A creation of E. Kasner: a googol is 10 to the one hundreth power, and a googolplex is 10 to the googol power. See E. Kasner and J. Newman, *Mathematics and the Imagination* (New York: Simon and Schuster, 1940) 25.

[66] See my "The Number of English Sentences", *Foundations of Language* II (1974) 519–432, for a discussion of the problems to be encountered in pondering lengthy imponderable sentences.

[67] Kasner, *op. cit.* 246.

[68] After all, the truth is the other way round. But then, of course, I hear Josef muttering, *sotto voce*, 'The truth is still another turn around'. You (the reader) may conclude that I am wrong, but there's no way that you can conclude that he's right.

[69] A somewhat similar point was made by Malcolm: "One statement about physical things *turned out to be false* only because you *made sure* of another statement about physical things". *Op. cit.* 69.

[70] From this standpoint, one can see that Malcolm's claim that "It impossible that *every* statement about physical things *could* turn out to be false", *op. cit.*, 69, is weaker than it may look; for nothing precludes the possibility that such statements may turn out to be profound conceptual confusions, and hence devoid of truth value.

[71] Here it may help to recall the identification of Grunt in Chapter III, above. In the expression 'Grunt is the giraffe who lived in my garden in Nairobi', 'Grunt' is a name and 'the giraffe who lived in my garden in Nairobi' is a description; thus the identification indicated a characterization of Grunt.

[72] I have heard of an aphasic who, when shown a toothbrush and asked what it was, was unable to say; but when the toothbrush was removed from sight, said 'That was a toothbrush'; presumably there was some interference between the circuitry governing visual inputs and associated verbal behavior.

BIBLIOGRAPHY
(of works cited in the text)

A. J. Ayer, *The Problem of Knowledge* (New York: St. Martin's Press, 1965).

H. Bettenson, ed. *Documents of the Christian Church* (New York & London: Oxford University Press, 1947).

N. Bohr, *Atomic Physics And Human Knowledge*, (New York: Science Editions, Inc., 1961).

C. Cherry, *On Human Communication* (New York: John Wiley & Sons, Inc., 1957).

J. Cook-Wilson, *Statement And Inference* (Oxford: Clarendon Press, 1926).

T. Givón, *On Understanding Grammar* (New York: Academic Press, 1979).

N. Goodman, *Fact, Fiction and Forecast* (Indianapolis-New York: Bobbs-Merrill Co.,Inc., 1973.)

Z. Harris, *Mathematical Structures of Language* (New York: John Wiley & Sons, 1968).

J. Hintikka, *Models of Modalities* (Dordrecht: D. Reidel, 1969).

E. Kasner, and J. Newman, *Mathematics and the Imagination* (New York: Simon And Schuster, 1940).

N. Malcolm, *Knowledge and Certainty* (Englewood Cliffs, N.J.: Prentice-Hall, Inc., 1963).

J. R. Newman, *The World Of Mathematics* (New York: Simon And Schuster, 1956).

J. R. Newman and E. Kasner, *Mathematics and the Imagination* (New York: Simon And Schuster, 1940).

R. Nozick, *Philosophical Explanations* (Cambridge: Harvard University Press, 1981).

W. V. O. Quine and J. S. Ullian, *The Web of Belief* (New York: Random House, 1970).

B. Russell, *The Problems of Philosophy* (New York: Henry Holt And Co.).

G. Ryle, *The Concept of Mind* (London: Hutchinson's University library, 1949).

E. Sapir, *Language* (New York; Harcourt Brace and Co., 1949)

J. Searle, *Speech* Acts (Cambridge: Cambridge University Press, 1970).

J. S. Ullian and W. V. O. Quine, *The Web of Belief* (New York: Random House, 1970).

P. Ziff, "Art and Sociobiology", *Mind* (1981) XC 505-520.

... *Philosophic Turnings* (Ithaca: Cornell University Press, 1966).

... *Semantic Analysis* (Ithaca: Cornell University Press, 1960).

... "The Number of English Sentences", *Foundations of Language* II (1974).

... *Understanding Understanding* (Ithaca and London: Cornell University Press, 1972).

INDEX